# ELECTROMAGNETISM

# ELECTROMAGNETIC SPECTRUM

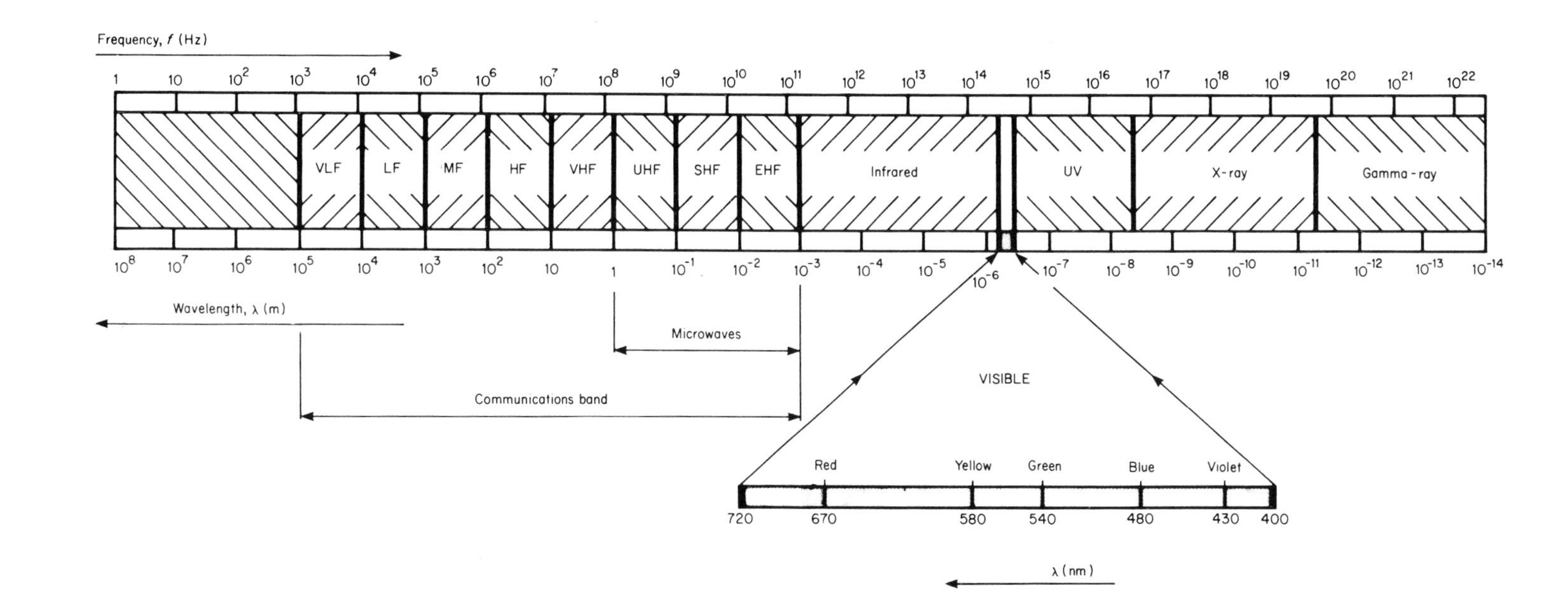

# ELECTROMAGNETISM

**V. ROSSITER**

*Trinity College*

*University of Dublin*

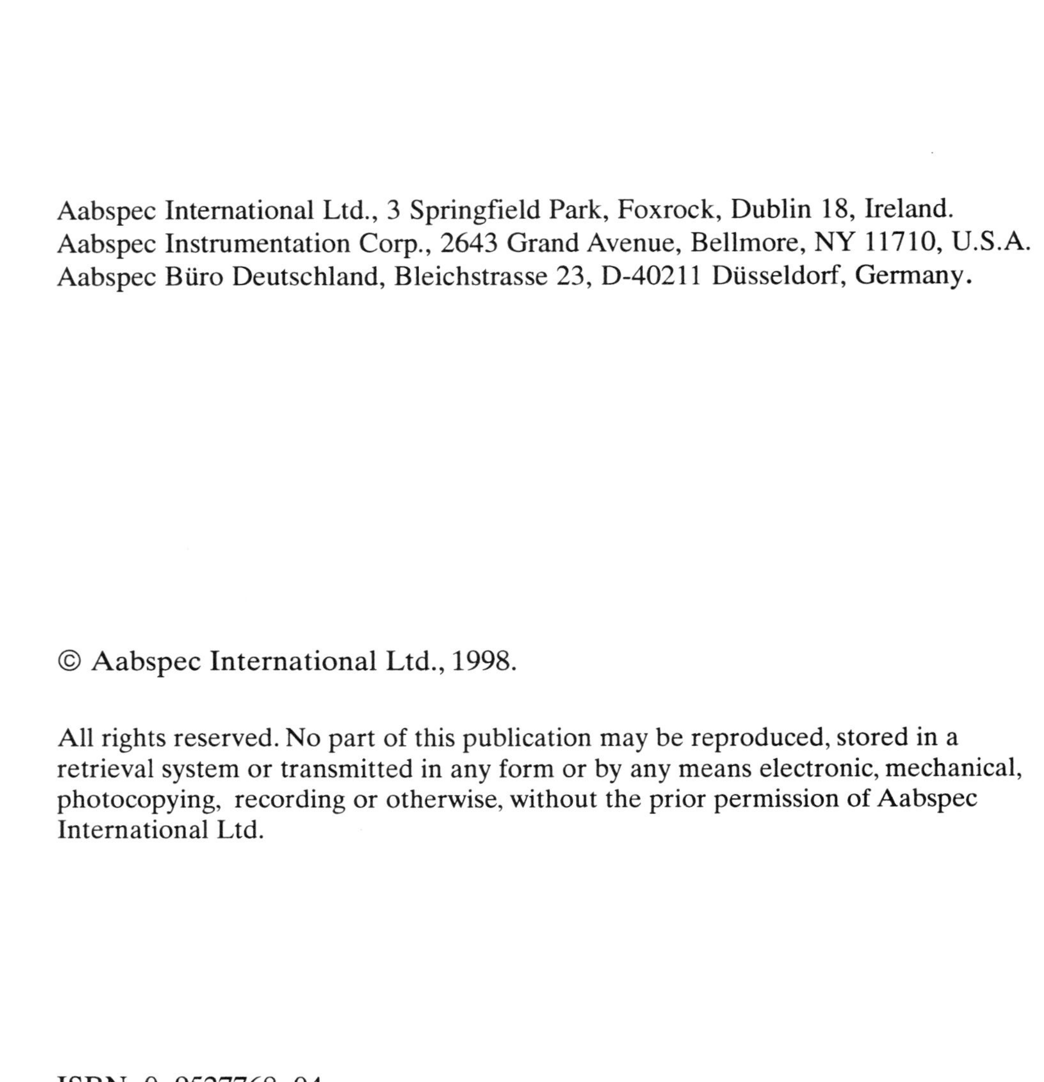

Aabspec International Ltd., 3 Springfield Park, Foxrock, Dublin 18, Ireland.
Aabspec Instrumentation Corp., 2643 Grand Avenue, Bellmore, NY 11710, U.S.A.
Aabspec Büro Deutschland, Bleichstrasse 23, D-40211 Düsseldorf, Germany.



ISBN 0 9527768 04

First edition published by Heyden & Sons Ltd.
This 1998 edition published by Aabspec International Ltd.

Printed in Ireland by Cahill Printers Ltd., Dublin.

# CONTENTS

Frontispiece ii

Preface ix

Acknowledgements xi

Chapter 1 ELECTRIC AND MAGNETIC FIELDS 1

1.1 Electric and Magnetic Forces: Lorentz Relationship 1
1.2 The Maxwell Equations 3
1.3 Some Important Mathematical Results 5
Exercises 5
1.4 Waves 10
Exercise 12
1.5 Plane Waves 13
1.6 The Three Dimensional Wave Equation 13
1.7 Energy in an Electromagnetic Field 14
Problems 17
Further Reading 17

Chapter 2 MAXWELL'S EQUATIONS IN FREE SPACE 18

2.1 Waves of **E** 18
2.2 Waves of **B** 19
2.3 Plane Electromagnetic Waves in Free Space 20
Exercises 26
2.4 Summary of Results 28
Problems 28

Chapter 3 MAXWELL'S EQUATIONS IN MATERIALS 29

3.1 A Model for the Individual Molecule 29
3.2 The Polarization Vector **P** 31
3.3 Our Molecular Model for a Material and **P** 34
3.4 The Electric Field within a Dense Material 35
3.5 Refractive Index and Relative Permittivity 38
Exercise 39
3.6 Summary of Results 41

Chapter 4 ELECTROMAGNETIC WAVES IN NON-CONDUCTING MATERIALS 43

4.1 Electromagnetic Waves in a Low Density Gas 43
4.2 Electromagnetic Waves in a Dense Material 46
4.3 The Consequences of a Complex Refractive Index 48
4.4 Phase Velocity and Energy Velocity 49
Exercise 50
4.5 Anomalous Dispersion 52
4.6 Group Velocity 52
Exercise 54
4.7 Summary of Results 54
Problems 55
Further Reading 55

Chapter 5 ELECTROMAGNETIC WAVES IN CONDUCTING MATERIALS 56

5.1 A Simple Model for a Metal 56
5.2 Metals at High and Low Frequencies 58
5.3 Reflection from a Plasma 62
Exercise 62
5.4 Guided Waves 65
5.5 Summary of Results 73
Appendix 5.1 74
Problems 74
Further Reading 75

Chapter 6 REFLECTION AND REFRACTION 76

6.1 The Boundary Conditions 76
6.2 The Propagation Vector 80
6.3 Reflection and Refraction at a Plane Boundary 81
6.4 Polarization by Reflection 93
6.5 Normal Incidence 93
6.6 Phase Change on Reflection 94
6.7 Real Materials 98
6.8 Summary of Results 98
Appendix 6.1 100
Problems 101
Further Reading 102

Chapter 7 THE SCALAR POTENTIAL AND THE VECTOR POTENTIAL 103

7.1 The Vector Potential **A** 103
7.2 The Scalar Potential $\phi$ 103
7.3 The Inhomogeneous Wave Equations in $\phi$ and **A** 104
7.4 Solutions for $\phi$ and **A** in Terms of the Field Sources $\rho$ and **J** 105
7.5 The Lienard-Wiechert Potentials 108
Appendix 7.1 114

Chapter 8 SIMPLE RADIATING SYSTEMS 116

8.1 Radiation from a Slowly Moving Accelerated Point Charge 116
8.2 Energy Scattered by a Free Charge 121
8.3 Scattering of Radiation by a Bound Charge 124
8.4 Radiation from an Electric Dipole Antenna 126
8.5 Radiation Resistance of an Antenna 128
8.6 Gain of an Antenna 128
8.7 Radiation from a Magnetic Dipole Antenna 129
8.8 Limitations in our Treatment of Antennas 131
Appendix 8.1 131
Appendix 8.2 133
Problems 138
Further Reading 139

Chapter 9 POLAR MATERIALS 140

9.1 Electromagnetic Energy in a Dielectric Material 140
9.2 Solids that are not always Solid 144
9.3 Ferroelectric and Ferromagnetic Materials 144
Further Reading 146

Appendix A REFERENCES TO SOURCES OF FURTHER READING 147

Appendix B SOME USEFUL MATHEMATICAL RESULTS 149

Appendix C THE RELATIONSHIPS BETWEEN **D** AND **H** AND VECTORS **E**, **B**, **P** AND **M** 156

Appendix D LIST OF SYMBOLS 158

Appendix E NOTES ON SI UNITS 161

Appendix F PHYSICAL CONSTANTS 162

Answers to Problems 163

Index 165

# PREFACE

This book has been written to provide a concise and simple outline of electromagnetism. The content is sufficient for a general purpose undergraduate course and the text will provide the necessary basis for more specialized and advanced applications. A knowledge of physics and mathematics to about first year university level is assumed. Students will generally find that they are able to work independently through the text which provides an essentially self-contained course on electromagnetism. The text is intended to be of interest to students of Engineering, Physics, Applied Mathematics and Chemistry.

Electromagnetism is a subject which is generally acknowledged as presenting particular difficulties for both teachers and students. Many of the well established texts treat the subject at a very sophisticated level and, although they are commendable in their own right, such an approach can be very off-putting for those new to the subject. On the other hand, the subject requires to be treated with reasonable mathematical rigour. The problem is to find a way of presenting the subject so that it is digestible while remaining consistent and logical through the unity of the Maxwell equations. This text is therefore different in style and format from the many other texts written on electromagnetism. The Maxwell equations are introduced and used throughout in terms of the field sources and the **E** and **B** fields only. This avoids the difficulties which can arise through the early introduction and definition of the additional field vectors **D** and **H**. Such a formulation of the Maxwell equations is not, in itself, new. It has formed the basis of Feynman's very beautiful 'Lectures on Physics' and indeed Volume 2 of that text is an ideal source of additional reading on many topics. The present text introduces the Maxwell equations at an early stage in Chapter 1. The equations are then applied in a variety of physical situations and the important applications are developed in a related and structured manner. Sufficient detail is provided to give a thorough treatment of each topic to the extent to which it is developed and to show the interchange of physical ideas and their mathematical description. Some of the topics are developed as exercises to encourage the reader to become actively involved. The exercises are by no means trivial and it is not always expected that the reader will be successful. For this reason, fully worked solutions are generally given. The numerical problems at the ends of the chapters have been selected to illustrate the text and to develop a feeling for relative magnitudes in various physical situations. The answers are given for numerical problems and SI units (Système International d'Unités) are used throughout.

In choosing the topics which are included, the general policy has been to treat only those topics which can be handled with reasonable mathematical rigour while using only commonly available mathematical skills. Only subjects which can be treated on a classical basis have been included. Obviously, in a small text of this type, many of the topics can be little more than introduced. Specialist students will need to develop topics in their particular areas to a more detailed and advanced level. Specific references to sources of further reading are given at the end of each chapter and the general list of references includes many advanced and specialized texts.

Val Rossiter
Trinity College, University of Dublin
April 1979

# ACKNOWLEDGEMENTS

I am indebted to my colleagues and students of the past few years. The text was developed as part of a course option in electromagnetic theory-communications for final year engineering students, and the students taking this subject have each contributed to the development of the present text with their many helpful suggestions. I very much appreciate the assistance given by Dr Tom Ambrose, Department of Mathematics, College of Technology, Kevin Street, Dublin 8, who read the manuscript during its early development. My thanks are due to Professor Cyril Delaney, Physics Department, Trinity College, Dublin, for some very stimulating and enjoyable discussions of reflection and refraction. There are many others who have been kind enough to read the manuscript and to make useful comments and to all those who provided such support I am indebted. It is with pleasure that I acknowledge the care and interest of Dr David Rees, Engineering School, Trinity College, Dublin, in the preparation of the illustrations.

*To my father,*

*Richard Rossiter*

*and in memory of*

*Richard Feynman.*

## Chapter 1

# ELECTRIC AND MAGNETIC FIELDS

We start by reviewing the forces which act on electric charges: the electric force described by the electric field **E** and the magnetic force described by the magnetic field **B**. The total electromagnetic force is given by the Lorentz relationship. This relationship effectively defines the electric field **E** and the magnetic field **B**. The Maxwell equations can then be written in their differential form in terms of the **E** and **B** fields. After a brief consideration of the experimental results on which the Maxwell equations are based, we turn our attention to the mathematical description of 'wave propagation'. This completes the base from which we can start to explore the information contained in the Maxwell equations.

## 1.1 ELECTRIC AND MAGNETIC FORCES: THE LORENTZ RELATIONSHIP

Experiment shows that there are two sets of circumstances under which an electric charge can experience a force due to another electric charge or group of charges.

The first case is where all the electric charges are stationary. Then each charge experiences an *electric force* due to the presence of the other charges. In the case of two small electric charges (point charges), the force acting on each charge is given by Coulomb's law. This law says that the force on a charge acts along the straight line joining the point charges and that the magnitude of the force varies inversely with the square of the distance separating the charges. We can write Coulomb's law in a neat mathematical form. Consider the force on a charge $q$ due to a charge $q_1$. Suppose we locate charge $q$ in relation to $q_1$ by means of a vector **R** drawn from $q_1$ as shown in Fig. 1.1.

Coulomb's law states that the force $\mathbf{F}_E$ experienced by $q$ is given by

$$\mathbf{F}_E = \frac{qq_1}{R^2}\left(\frac{\mathbf{R}}{R}\right)\left(\frac{1}{4\pi\varepsilon_0}\right) \tag{1.1}$$

We have incorporated the direction of $\mathbf{F}_E$ into Coulomb's law by means of the unit vector $\mathbf{R}/R$. This vector has a magnitude of unity and so makes no difference to the magnitude of the expression giving the force. In SI units the charges $q_1$

and $q$ are in coulomb's (C), the distance $R$ is in metres (m) and the magnitude of the force $F_E$ is in newtons (N). In this system of units, the Coulomb expression contains the constant term $1/4\pi\varepsilon_0$, where $\varepsilon_0$ is a constant known as the absolute permittivity of free space and $\varepsilon_0 = 8.854 \times 10^{-12}$ F m$^{-1}$ (farads per metre).

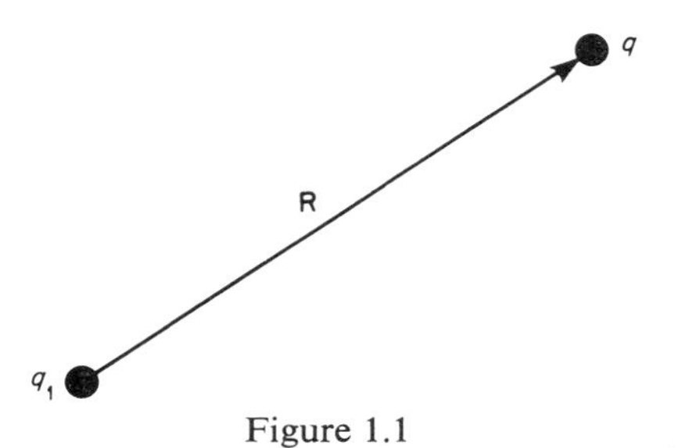


Figure 1.1

The information contained in Coulomb's law can be presented in a slightly different way. Instead of dealing with the force produced on $q$ by the charge $q_1$ we could consider the force produced per unit charge by $q_1$, that is, the value of $\mathbf{F}_E$ given by setting $q = 1$ C. This particular value of the electric force we can call the electric field $\mathbf{E}$

$$\mathbf{E} = \frac{q_1}{R^2}\left(\frac{\mathbf{R}}{R}\right)\left(\frac{1}{4\pi\varepsilon_0}\right) \tag{1.2}$$

The force on the charge $q$ can then be found by multiplying $q$ by $\mathbf{E}$, that is,

$$\mathbf{F}_E = q\mathbf{E} = \frac{qq_1}{R^2}\left(\frac{\mathbf{R}}{R}\right)\left(\frac{1}{4\pi\varepsilon_0}\right) \tag{1.3}$$

and we are back again to Coulomb's law.

The electric field $\mathbf{E}$ gives the force which acts between charges or groups of charges which are not in motion relative to one another. But if we have charges which are in relative motion then the situation is quite different and we find that there is another type of force under these circumstances. This type of force can be observed to act between two current carrying wires. In this case the force is between charges which are in motion, that is, between currents. We call this force the *magnetic force*. Even though the situation is now more complicated, we can define a magnetic field $\mathbf{B}$ which is the force experienced by a unit current in the presence of another current. For a charge $q$ moving with a velocity $\mathbf{v}$ in a magnetic field, the magnetic force $\mathbf{F}_B$ experienced by the charge is given by

$$\mathbf{F}_B = q\mathbf{v} \times \mathbf{B} \tag{1.4}$$

Naturally there is an expression which gives the magnetic field $\mathbf{B}$ in terms of its current sources: just as Eqn (1.2) gives the $\mathbf{E}$ field in terms of its source, the electric charge. We leave until later the determination of the $\mathbf{B}$ field from a description of the currents producing it. For the moment we are really concerned with the effects of **E** and **B**—indeed we can define **E** and **B** on just this basis. The total force experienced by a charge due to the combined electric and magnetic forces is given by

$$\mathbf{F} = \mathbf{F}_E + \mathbf{F}_B$$

that is

$$\boxed{\mathbf{F} = q\mathbf{E} + q\mathbf{v} \times \mathbf{B}} \tag{1.5}$$

This is the expression for the Lorentz force and we can take this expression as the definition of **E** and **B**.

In terms of Eqn (1.5), the units of the electric field **E** are those of force per unit charge or newtons per coulomb, that is, $\mathrm{N\,C^{-1}}$. In the SI system this is equivalent to volts per metre, that is $\mathrm{V\,m^{-1}}$. Similarly the units of the magnetic field **B** are those of force per unit charge per unit velocity, or $\mathrm{N\,C^{-1}\,m^{-1}\,s}$. In the SI system this unit is known as the tesla (T).

## 1.2 THE MAXWELL EQUATIONS

There are only four Maxwell equations and between them they contain only four variables. At first they may appear a little strange but, because there are so few terms, it does not take long to become familiar with them. Let's write them out in the form in which we intend to use them:

$$\nabla \cdot \mathbf{E} = \rho/\varepsilon_0 \tag{1.6}$$

$$\nabla \times \mathbf{E} = -\partial \mathbf{B}/\partial t \tag{1.7}$$

$$\nabla \cdot \mathbf{B} = 0 \tag{1.8}$$

$$c^2 \nabla \times \mathbf{B} = \mathbf{J}/\varepsilon_0 + \partial \mathbf{E}/\partial t \tag{1.9}$$

The electric field vector **E** and the magnetic field vector **B** have the meaning given in the expression for the Lorentz force. The constant $\varepsilon_0$ has been given above and the constant $c$ is the velocity of light in free space, equal to (approximately) $3 \times 10^8\ \mathrm{ms^{-1}}$. The mathematical operator del $\nabla$ ($\equiv \mathbf{i}\,\partial/\partial x + \mathbf{j}\,\partial/\partial y + \mathbf{k}\,\partial/\partial z$ in rectangular coordinates) and the partial derivative with respect to time, have their usual meaning.

Equation (1.6) relates the electric field **E** to the charge density $\rho$ ($\mathrm{C\,m^{-3}}$) where the function $\rho$ allows us to treat the distribution of charge in space in a very general way. Similarly, in Eqn (1.9), the magnetic field **B** is related to the existence of electric currents which can be described by the current density vector **J** ($\mathrm{A\,m^{-2}}$). In general, each of the terms **E**, **B**, $\rho$ and **J** are functions of space and time so that we can follow, for example, the changing **E** and **B** fields produced at a point in space as a charge passes by. A point in space is described by the vector **r** drawn from some reference point (the origin) to the point of interest (the field point). A position in time is described by the time coordinate $t$. Thus we can completely specify any point in terms of coordinates $(\mathbf{r}, t)$. The variation of $\mathbf{E}$, $\mathbf{B}$, $\rho$ and $\mathbf{J}$ from point to point will generally be left as 'understood', but if we want to emphasize the fact we can write the quantities as functions of space and time:

$$\begin{aligned} \nabla \cdot \mathbf{E}(\mathbf{r}, t) &= \rho(\mathbf{r}, t)/\varepsilon_0 \\ \nabla \times \mathbf{E}(\mathbf{r}, t) &= -\partial \mathbf{B}(\mathbf{r}, t)/\partial t \\ \nabla \cdot \mathbf{B}(\mathbf{r}, t) &= 0 \\ c^2 \nabla \times \mathbf{B}(\mathbf{r}, t) &= \mathbf{J}(\mathbf{r}, t)/\varepsilon_0 + \partial \mathbf{E}(\mathbf{r}, t)/\partial t \end{aligned}$$

Although we haven't written the equations in the same form that Maxwell used, we still take them for what they were when Maxwell wrote them—generalizations of experimental results. We can now have a preliminary discussion of these equations and identify each with the experimental result on which it is based.

The first equation

$$\nabla \cdot \mathbf{E} = \rho/\varepsilon_0$$

relates the electric field and the source of such fields—an electric charge density. This equation is in fact a form of Coulomb's law, as we will later show.

The second equation

$$\nabla \times \mathbf{E} = -\partial \mathbf{B}/\partial t$$

suggests that a magnetic field which changes with time will give rise to an electric field. We know that if we set up a magnetic field near a loop of wire and quickly reduce the magnetic field to zero, the result is that an electric field is produced which causes the electrons in the wire to move to give an induced current. This equation is a differential form of Faraday's law of electromagnetic induction.

The third equation

$$\nabla \cdot \mathbf{B} = 0$$

makes an interesting comparison with the first equation. The divergence of **B** is not at all like the divergence of **E**. This indicates that there is a difference in type between the sources of **E** and **B**. Whereas the sources of **E** are *point* charges this cannot be the case for **B**—in other words we are being told that there are no 'point magnetic charges' or 'free magnetic poles' to act as point sources for the magnetic field. Despite the search (which has taken place for some years now) it remains a fact that such free magnetic poles have not been found.

The fourth equation

$$c^2 \nabla \times \mathbf{B} = \mathbf{J}/\varepsilon_0 + \partial \mathbf{E}/\partial t$$

again stresses the intimate connection between the electric and magnetic fields and suggests that it is current (or moving charge) which serves as the source of the magnetic field. The equation is an extension of Ampère's law which describes the magnetic field produced by a current.

## 1.3 SOME IMPORTANT MATHEMATICAL RESULTS

It is obvious from the way in which we have written the Maxwell equations that, if we are to unravel the information they contain, then we will need to use some vector analysis and vector calculus. A more complete guide to the mathematical background is given in Appendix B but some of the most important results are given here.

### Vector Identities

For any arbitrary vector $\mathbf{C}$ the following relations hold

$$\nabla \times (\nabla \times \mathbf{C}) = \nabla(\nabla \cdot \mathbf{C}) - \nabla^2 \mathbf{C} \tag{1.10}$$

$$\nabla \cdot (\nabla \times \mathbf{C}) = 0 \tag{1.11}$$

### Vector Integrals

For any arbitrary vector $\mathbf{C}$ the following vector integrals hold

(i) Gauss' Theorem

$$\int_s \mathbf{C} \cdot \mathbf{n}\, \mathrm{d}s = \int_V \nabla \cdot \mathbf{C}\, \mathrm{d}V \tag{1.12}$$

where the volume $V$ is enclosed by the surface $s$ and where $\mathbf{n}$ is the outward unit normal to $s$ at an element of surface area $\mathrm{d}s$.

(ii) Stokes' Theorem

$$\oint_\Gamma \mathbf{C} \cdot \mathrm{d}\mathbf{r} = \int_s (\nabla \times \mathbf{C}) \cdot \mathbf{n}\, \mathrm{d}s \tag{1.13}$$

where $\mathrm{d}\mathbf{r}$ is an element of the closed curve $\Gamma$ which forms the boundary of the surface $s$. Again, $\mathbf{n}$ is the unit normal to $s$ at an element of surface area $\mathrm{d}s$.

## EXERCISE 1.1

Use Gauss' Integral Theorem to obtain Coulomb's law for the force between two point charges from the Maxwell equation $\nabla \cdot \mathbf{E} = \rho/\varepsilon_0$.

### Solution

We start by finding the electric field due to a point charge. The point charge can be regarded as the limiting case (as the volume approaches zero) of a charge distribution. Take our point charge $q_1$ to be at the origin and surrounded by an imaginary sphere of radius $R$ as shown in Fig. 1.2.

Then the charge $q_1$ is accounted for if we integrate the charge density $\rho$ over the volume of the sphere

$$q_1 = \int_V \rho \, \mathrm{d}V$$

Therefore, taking $\rho$ from

$$\nabla \cdot \mathbf{E} = \rho/\varepsilon_0$$

we can write

$$q_1/\varepsilon_0 = \int_V \nabla \cdot \mathbf{E} \, \mathrm{d}V$$

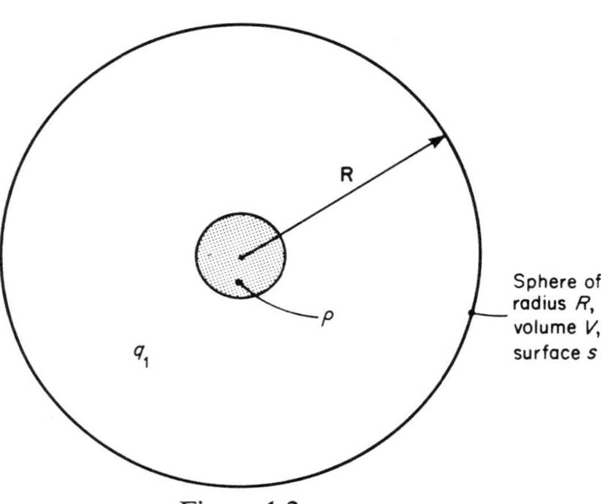


Figure 1.2

The volume integral is transformed into the corresponding surface integral by Gauss' theorem

$$q_1/\varepsilon_0 = \int_s \mathbf{E} \cdot \mathbf{n} \, \mathrm{d}s$$

The symmetry of the problem (with the charge at the centre of the sphere there is no reason to assume that any point on the surface will be different from any other) suggests that the value of $\mathbf{E} \cdot \mathbf{n}$ must be the same at all points on the sphere. Therefore

$$q_1/\varepsilon_0 = \mathbf{E} \cdot \mathbf{n} \int_s \mathrm{d}s$$

$$= \mathbf{E} \cdot \mathbf{n} \, (4\pi R^2)$$

where the integral of d$s$ simply gives the surface area of the sphere. That is

$$\mathbf{E} \cdot \mathbf{n} = \frac{q_1}{R^2} \left( \frac{1}{4\pi\varepsilon_0} \right)$$

or

$$\mathbf{E} = \frac{q_1}{R^2}\left(\frac{\mathbf{R}}{R}\right)\left(\frac{1}{4\pi\varepsilon_0}\right)$$

since $\mathbf{R}/R$ is a unit vector in the direction of $\mathbf{n}$, that is, $\mathbf{R}/R \equiv \mathbf{n}$.

The electric force experienced by a charge $q$ in the vicinity of $q_1$ is given by

$$\mathbf{F}_E = q\mathbf{E} = \frac{qq_1}{R^2}\left(\frac{\mathbf{R}}{R}\right)\left(\frac{1}{4\pi\varepsilon_0}\right)$$

which is Coulomb's law.

## EXERCISE 1.2

An electric dipole is formed by separating the charges $+q$ and $-q$ by a small distance $R_0$. The vector $\mathbf{R}_0$ is taken in the direction of movement of the positive charge from the neutral condition before the charges are separated. The distance $R_0$ is much smaller than the distance $R$ where $\mathbf{R}$ is drawn from the dipole to the field point. From the expression for the electric field due to a point charge

$$\mathbf{E} = \frac{q}{R^2}\left(\frac{\mathbf{R}}{R}\right)\left(\frac{1}{4\pi\varepsilon_0}\right)$$

show that the field due to the dipole is given by

$$\left(\frac{1}{4\pi\varepsilon_0}\right)\left[\frac{3R_0 q\cos\theta}{R^3}\left(\frac{\mathbf{R}}{R}\right) - \frac{q\mathbf{R}_0}{R^3}\right]$$

where $\theta$ is the angle between $\mathbf{R}_0$ and $\mathbf{R}$. If the 'electric dipole moment' $\mathbf{p}$ is defined as $q\mathbf{R}_0$, show that the expression for the field can be written as

$$\left(\frac{1}{4\pi\varepsilon_0}\right)\left[\frac{3(\mathbf{p}\cdot\mathbf{R})\mathbf{R}}{R^5} - \frac{\mathbf{p}}{R^3}\right]$$

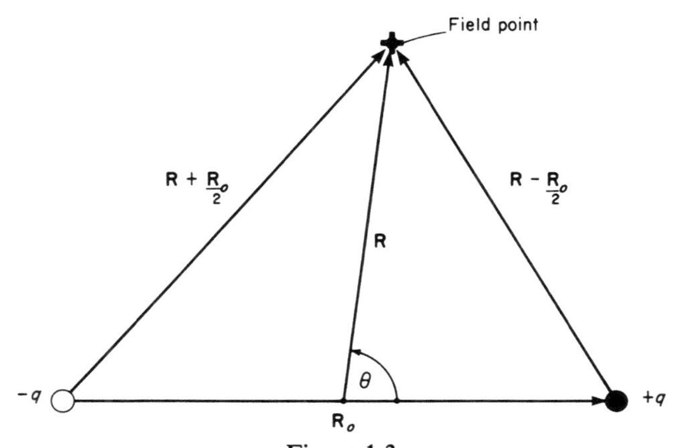


Figure 1.3

**Solution**

The electric field at the field point is given by

$$\mathbf{E} = \left(\frac{1}{4\pi\varepsilon_0}\right)\left\{\frac{q}{|\mathbf{R}-\mathbf{R}_0/2|^2}\frac{(\mathbf{R}-\mathbf{R}_0/2)}{|\mathbf{R}-\mathbf{R}_0/2|} - \frac{q}{|\mathbf{R}+\mathbf{R}_0/2|^2}\frac{(\mathbf{R}+\mathbf{R}_0/2)}{|\mathbf{R}+\mathbf{R}_0/2|}\right\}$$

where we are considering the static case only. The distances from the charges to the field point are obtained from the cosine rule as

$$|\mathbf{R}-\mathbf{R}_0/2| = \left[R^2+\left(\frac{R_0}{2}\right)^2-2R\left(\frac{R_0}{2}\right)\cos\theta\right]^{\frac{1}{2}}$$

$$|\mathbf{R}+\mathbf{R}_0/2| = \left[R^2+\left(\frac{R_0}{2}\right)^2-2R\left(\frac{R_0}{2}\right)\cos(180^\circ-\theta)\right]^{\frac{1}{2}}$$

which for $R \gg R_0$ is

$$|\mathbf{R}-\mathbf{R}_0/2| = R\left[1-\frac{R_0}{R}\cos\theta\right]^{\frac{1}{2}}$$

$$|\mathbf{R}+\mathbf{R}_0/2| = R\left[1+\frac{R_0}{R}\cos\theta\right]^{\frac{1}{2}}$$

Therefore,

$$|\mathbf{R}-\mathbf{R}_0/2|^{-3} = \frac{1}{R^3}\left(1-\frac{R_0}{R}\cos\theta\right)^{-\frac{3}{2}} = \frac{1}{R^3}\left(1+\frac{3}{2}\frac{R_0}{R}\cos\theta+\cdots\right)$$

and since $(R_0/R) \ll 1$ the higher powers in $(R_0/R)$ can be neglected. Similarly we obtain

$$|\mathbf{R}+\mathbf{R}_0/2|^{-3} = \frac{1}{R^3}\left(1+\frac{R_0}{R}\cos\theta\right)^{-\frac{3}{2}} = \frac{1}{R^3}\left(1-\frac{3}{2}\frac{R_0}{R}\cos\theta+\cdots\right)$$

The electric field is then given by

$$\mathbf{E} = \left(\frac{1}{4\pi\varepsilon_0}\right)\left\{\frac{q\mathbf{R}}{R^3}\left[\left(1+\frac{3}{2}\frac{R_0}{R}\cos\theta\right)-\left(1-\frac{3}{2}\frac{R_0}{R}\cos\theta\right)\right]\right.$$
$$\left.-\frac{q\mathbf{R}_0}{2R^3}\left[\left(1+\frac{3}{2}\frac{R_0}{R}\cos\theta\right)+\left(1-\frac{3}{2}\frac{R_0}{R}\cos\theta\right)\right]\right\}$$
$$= \left(\frac{1}{4\pi\varepsilon_0}\right)\left\{\frac{3R_0 q\cos\theta}{R^3}\left(\frac{\mathbf{R}}{R}\right)-\frac{q\mathbf{R}_0}{R^3}\right\}$$

which is the required result.

If we define the electric dipole moment by

$$\mathbf{p} = q\mathbf{R}_0$$

then

$$R_0 q \cos\theta = p\cos\theta$$
$$= \frac{\mathbf{p}\cdot\mathbf{R}}{R}$$

and we can write the field as

$$\mathbf{E} = \left(\frac{1}{4\pi\varepsilon_0}\right)\left\{\frac{3(\mathbf{p}\cdot\mathbf{R})\mathbf{R}}{R^5} - \frac{\mathbf{p}}{R^3}\right\}$$

which is the result given above.

Now see if you can show that if we take $\mathbf{r}_p$ as being drawn from the field point to the dipole, then the field is given by

$$\mathbf{E} = \left(\frac{1}{4\pi\varepsilon_0}\right)\left\{\frac{3(\mathbf{p}\cdot\mathbf{r}_p)\mathbf{r}_p}{r_p^5} - \frac{\mathbf{p}}{r_p^3}\right\}$$

The working is essentially the same as that given above except that now $\mathbf{r}_p$ is in the opposite direction to $\mathbf{R}$.

*Note:*
To obtain the above results for the field due to a dipole, we have made use of the 'principle of superposition': that is, that the field at a point due to a number of charges is given by the vector sum of the fields at that point due to each individual charge.

## EXERCISE 1.3

Neglecting edge effects, show that the static electric field $E$ between the plates of a parallel plate capacitor is given by

$$E = \sigma/\varepsilon_0$$

where $\sigma$ is the charge per unit area on the positive plate.

### Solution

It is clear that the electric field between the plates does not have any component which is not perpendicular to the surface of the plate; this is because any parallel component would induce charge flow in the metal until its value was zero. Consequently the electric field is perpendicular to the positive plate and in the direction from the positive to the negative plate. We now construct an imaginary unit cube so that it contains unit area of the surface of the positive plate and with one of its faces parallel to the plate surface.

The problem now becomes very similar to that discussed in Exercise 1.1. The charge $Q$ within the unit volume is given by

$$Q = \int_V \rho \, \mathrm{d}V$$
$$= \sigma \text{ (numerically)}$$

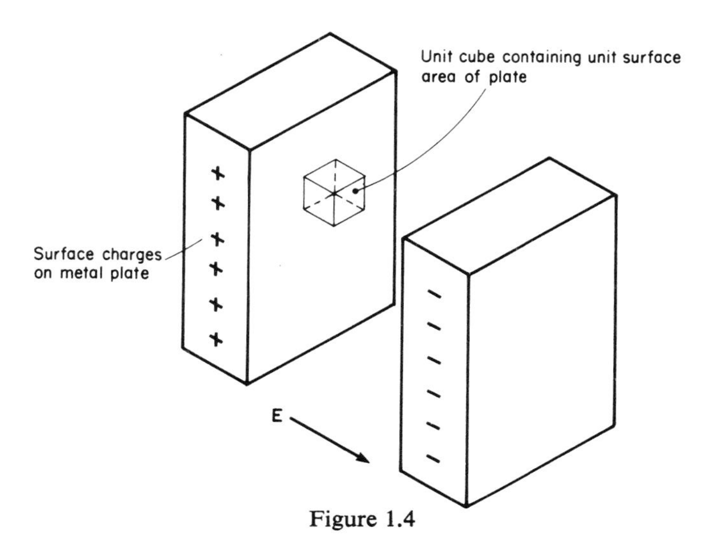


Figure 1.4

since the total charge contained in $V$ is on unit area of the positive plate. The Maxwell relation

$$\nabla \cdot \mathbf{E} = \rho/\varepsilon_0$$

gives

$$\sigma/\varepsilon_0 = \int_V \nabla \cdot \mathbf{E} \, \mathrm{d}V$$

which from Gauss' Theorem is

$$\sigma/\varepsilon_0 = \int_s \mathbf{E} \cdot \mathbf{n} \, \mathrm{d}s$$

The term $\mathbf{E} \cdot \mathbf{n}$ is non-zero over only one face of the unit cube at which stage $\mathbf{E}$ and $\mathbf{n}$ are in the same direction: therefore

$$\sigma/\varepsilon_0 = E$$

which is the required result.

## 1.4 WAVES

It is a matter of everyday experience (radio, light, etc.) that electromagnetic waves can propagate through free space and through various materials. The behaviour of these waves is described by Maxwell's equations but before we can begin to study such phenomena we need to know how to describe a wave in mathematical terms. We start by satisfying ourselves that *any* function of $(z-vt)$ represents a wave travelling in the $z$-direction with velocity $v$. Suppose that at time zero ($t = 0$) the function looks like that shown in Fig. 1.5.

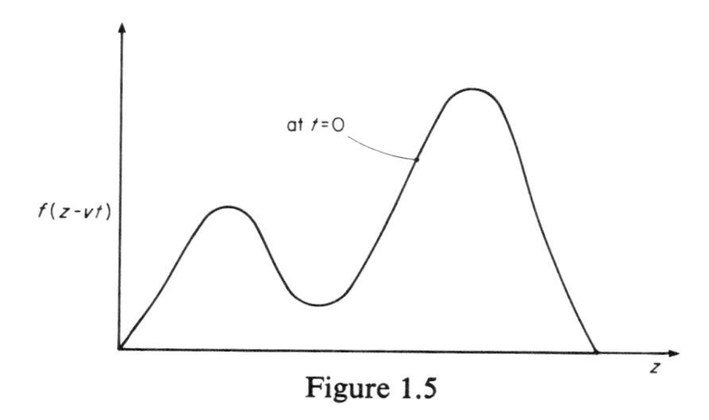


Figure 1.5

We have described how the function varies in space (with $z$) at time zero. Can we now decide how it will vary in space when the time $t = 1$ has been reached? If we substitute $t = 1$ into the variable $(z-vt)$ we obtain $(z-v)$, so in advancing the time we have reduced the variable by $v$. Plotting the function at $t = 1$ is now very simple since the value of the function at each value of $z$ corresponds to the function at $t = 0$ at the point $z-v$.

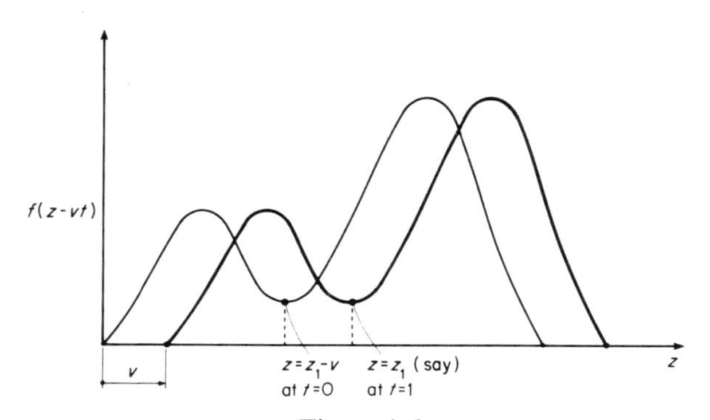


Figure 1.6

The result is that the entire picture moves (see Fig. 1.6) forward along the $z$-axis by an amount $v$ as the time goes from $t = 0$ to $t = 1$. We have then a 'wave' moving in the positive $z$-direction with a velocity $v$. Any function of $(z-vt)$ will do this. The 'function' simply determines the form of the wave in space at any time. This form is then carried forward with velocity $v$. Similarly, any function of $(z+vt)$ represents a wave travelling in the negative $z$-direction with time.

Now we can show that functions of $(z+vt)$ and $(z-vt)$ satisfy the equation known as the one dimensional wave equation, that is,

$$\frac{\partial^2\psi}{\partial z^2} = \frac{1}{v^2}\frac{\partial^2\psi}{\partial t^2} \qquad 1.14$$

In the one dimensional wave equation, $v$ represents the velocity of propagation of the quantity $\psi$. The term $\psi$ can represent anything which might vary in space and time, for example, force, displacement or probability.

## EXERCISE 1.4

Show that $\psi = f(z-vt)+g(z+vt)$ will satisfy the one dimensional wave equation.

### Solution

Consider each of the two functions in $\psi$ separately and write

$$\psi_f = f(z-vt)$$

Differentiation with respect to $z$ gives

$$\frac{\partial\{f(z-vt)\}}{\partial z} = f'(z-vt)$$

where we use the abbreviation

$$f'(z-vt) \quad \text{for} \quad \frac{\partial\{f(z-vt)\}}{\partial(z-vt)}$$

and similarly $f''$ denotes the second derivative with respect to $(z-vt)$. Hence,

$$\frac{\partial^2\psi_f}{\partial z^2} = f''(z-vt)$$

The time derivative is then

$$\frac{\partial}{\partial t}\{f(z-vt)\} = (-v)f'(z-vt)$$

and

$$\frac{\partial^2}{\partial t^2}\{f(z-vt)\} = v^2 f''(z-vt)$$

$$= \frac{\partial^2\psi_f}{\partial t^2}$$

Therefore

$$\frac{1}{v^2}\frac{\partial^2 \psi_f}{\partial t^2} = \frac{\partial^2 \psi_f}{\partial z^2}$$

which is the one dimensional wave equation. Similarly for $\psi_g = g(z+vt)$. If any two functions separately satisfy a differential equation then it follows that their sum must also satisfy the equation. Thus we show that

$$\psi = f(z-vt)+g(z+vt)$$

satisfies the one dimensional wave equation for any function of $(z-vt)$ and for any function of $(z+vt)$.

## 1.5 PLANE WAVES

Up to now our discussion of 'waves' has only involved one dimension, $z$. We can describe a quantity which varies with $z$ and travels in the $z$-direction. Now we broaden this restricted one dimensional wave to the simplest type of three dimensional wave, the plane wave. All we need to say is that whereas our wave only varies with $z$, it still has a value as we move about in the $x$-$y$ plane. So if we choose $t$ and $z$ we determine the value of the function throughout the corresponding $x$-$y$ plane. Although plane waves vary with $z$ and $t$ only, they do extend into all three space dimensions.

## 1.6 THE THREE DIMENSIONAL WAVE EQUATION

This is the equation

$$\frac{\partial^2 \psi}{\partial x^2}+\frac{\partial^2 \psi}{\partial y^2}+\frac{\partial^2 \psi}{\partial z^2} = \frac{1}{v^2}\frac{\partial^2 \psi}{\partial t^2}$$

which can be written as

$$\nabla^2 \psi = \frac{1}{v^2}\frac{\partial^2 \psi}{\partial t^2} \tag{1.15}$$

This equation is very similar to the one dimensional wave equation and the superposition of three one dimensional waves will satisfy the three dimensional wave equation. That is,

$$\psi = X(x-vt)+Y(y-vt)+Z(z-vt)$$

Then,

$$\nabla^2\psi = \left(\frac{\partial^2}{\partial x^2}+\frac{\partial^2}{\partial y^2}+\frac{\partial^2}{\partial z^2}\right)\psi$$

$$= \left(\frac{\partial^2}{\partial x^2}+\frac{\partial^2}{\partial y^2}+\frac{\partial^2}{\partial z^2}\right)\{X(x-vt)+Y(y-vt)+Z(z-vt)\}$$

$$= \frac{\partial^2 X(x-vt)}{\partial^2 x^2}+\frac{\partial^2 Y(y-vt)}{\partial^2 y^2}+\frac{\partial^2 Z(z-vt)}{\partial^2 z^2}$$

$$= X''(x-vt)+Y''(y-vt)+Z''(z-vt)$$

where the notation $X''$, $Y''$, $Z''$ is used to show the second derivative with respect to $(x-vt)$, $(y-vt)$ and $(z-vt)$, respectively.

Similarly,

$$\frac{\partial^2\psi}{\partial t^2} = v^2X''(x-vt)+v^2Y''(y-vt)+v^2Z''(z-vt)$$

and hence we have shown that the function satisfies the wave equation,

$$\nabla^2\psi = \frac{1}{v^2}\frac{\partial^2\psi}{\partial t^2}$$

## 1.7 ENERGY IN AN ELECTROMAGNETIC FIELD

The Maxwell equations lead to an equation for the conservation of electromagnetic energy. This equation contains a term which can be identified with the rate of flow of energy at a point in an electromagnetic field.

We have written the Maxwell equations as

$$\nabla\cdot\mathbf{E} = \rho/\varepsilon_0 \tag{1.16}$$

$$\nabla\times\mathbf{E} = -\partial\mathbf{B}/\partial t \tag{1.17}$$

$$\nabla\cdot\mathbf{B} = 0 \tag{1.18}$$

$$c^2\nabla\times\mathbf{B} = \mathbf{J}/\varepsilon_0+\partial\mathbf{E}/\partial t \tag{1.19}$$

Taking the scalar product of $\mathbf{E}$ with Eqn (1.19) gives

$$c^2\mathbf{E}\cdot\nabla\times\mathbf{B} = \mathbf{E}\cdot\mathbf{J}/\varepsilon_0+\mathbf{E}\cdot\partial\mathbf{E}/\partial t \tag{1.20}$$

The scalar product of $\mathbf{B}$ with Eqn (1.17) gives

$$\mathbf{B}\cdot\nabla\times\mathbf{E} = -\mathbf{B}\cdot\partial\mathbf{B}/\partial t$$

and multiplying by $c^2$ gives

$$c^2\mathbf{B}\cdot\nabla\times\mathbf{E} = -c^2\mathbf{B}\cdot\partial\mathbf{B}/\partial t \tag{1.21}$$

Subtracting this equation from Eqn (1.20) gives

$$c^2[\mathbf{E}\cdot\nabla\times\mathbf{B}-\mathbf{B}\cdot\nabla\times\mathbf{E}] = \mathbf{E}\cdot\mathbf{J}/\varepsilon_0+\mathbf{E}\cdot\partial\mathbf{E}/\partial t+c^2\mathbf{B}\cdot\partial\mathbf{B}/\partial t \tag{1.22}$$

On the left hand side we have a term containing two scalar triple products which we can combine provided we are careful about the operator $\nabla$. In the first of these products, the operation $\nabla\times$ is performed on $\mathbf{B}$ only, while in the second product the operation $\nabla\times$ affects $\mathbf{E}$ only. To make this quite clear let us write the term as

$$c^2[\mathbf{E}\cdot\nabla_B\times\mathbf{B}-\mathbf{B}\cdot\nabla_E\times\mathbf{E}]$$

For any three vectors $\mathbf{a}$, $\mathbf{b}$, $\mathbf{C}$ we can rotate the terms in their scalar triple product so that

$$\mathbf{a}\cdot\mathbf{b}\times\mathbf{C} = \mathbf{b}\cdot\mathbf{C}\times\mathbf{a}$$

This can be readily verified by multiplying out the expression and the general result is given in Appendix B. Using this result we can write

$$\begin{aligned} c^2[\mathbf{E}\cdot\nabla_B\times\mathbf{B}-\mathbf{B}\cdot\nabla_E\times\mathbf{E}] &= c^2[\nabla_B\cdot\mathbf{B}\times\mathbf{E}-\nabla_E\cdot\mathbf{E}\times\mathbf{B}] \\ &= c^2[-\nabla_B\cdot\mathbf{E}\times\mathbf{B}-\nabla_E\cdot\mathbf{E}\times\mathbf{B}] \\ &= -c^2[(\nabla_B+\nabla_E)\cdot\mathbf{E}\times\mathbf{B}] \\ &= -c^2\nabla\cdot\mathbf{E}\times\mathbf{B} \end{aligned}$$

Equation (1.22) may now be written

$$\varepsilon_0 c^2\nabla\cdot\mathbf{E}\times\mathbf{B} = -\mathbf{E}\cdot\mathbf{J}-\frac{\varepsilon_0}{2}\frac{\partial}{\partial t}(\mathbf{E}^2+c^2\mathbf{B}^2) \tag{1.23}$$

Now we must try to interpret the meaning of the terms in this equation. Consider the term $\mathbf{E}\cdot\mathbf{J}$. Suppose that the current density $\mathbf{J}$ is in the same direction as $\mathbf{E}$ and consists of $N$ charges, each of magnitude $q$, per unit volume moving with uniform velocity $\mathrm{d}x/\mathrm{d}t$.

Then,

$$\begin{aligned} \mathbf{E}\cdot\mathbf{J} &= ENq\frac{\mathrm{d}x}{\mathrm{d}t} \\ &= \frac{1}{\mathrm{d}t}(ENq\,\mathrm{d}x) \\ &= \frac{1}{\mathrm{d}t}\{N\times\text{Force on a charge}\times\text{an element of distance}\} \\ &= \frac{1}{\mathrm{d}t}\{\text{an element of work done}\} \end{aligned}$$

that is

$$\mathbf{E}\cdot\mathbf{J} = \frac{\mathrm{d}U}{\mathrm{d}t}$$

Therefore $\mathbf{E}\cdot\mathbf{J}$ can be identified as the rate of change of energy per unit volume at a point. If Eqn (1.23) is to be dimensionally correct, then all of the terms must have this meaning—the rate of change of energy per unit volume at a point in the field. Equation (1.23) is telling us about the energy balance which must be maintained. The term $\varepsilon_0 c^2 \mathbf{E}\times\mathbf{B}$ is known as the Poynting vector $\mathbf{S}$ and we can write the energy equation as

$$\nabla\cdot\mathbf{S} = -\mathbf{E}\cdot\mathbf{J} - \frac{\varepsilon_0}{2}\frac{\partial}{\partial t}(\mathbf{E}^2 + c^2\mathbf{B}^2) \tag{1.24}$$

To obtain a better understanding of $\mathbf{S}$ we can consider the equation in a region where the current density $\mathbf{J}$ will be zero, for example, in free space. Then,

$$\nabla\cdot\mathbf{S} = -\frac{\varepsilon_0}{2}\frac{\partial}{\partial t}(\mathbf{E}^2 + c^2\mathbf{B}^2) \tag{1.25}$$

Since we have decided that each term is the rate of change of energy per unit volume at a point in the field, we can find the rate of change of energy in any volume $V$ as

$$\int_V \nabla\cdot\mathbf{S}\,\mathrm{d}V = -\int_V \frac{\varepsilon_0}{2}\frac{\partial}{\partial t}(\mathbf{E}^2 + c^2\mathbf{B}^2)\,\mathrm{d}V \tag{1.26}$$

Applying Gauss' Integral Theorem to the left-hand side means we can write

$$\int_s \mathbf{S}\cdot\mathbf{n}\,\mathrm{d}s = -\frac{\varepsilon_0}{2}\int_V \frac{\partial}{\partial t}(\mathbf{E}^2 + c^2\mathbf{B}^2)\,\mathrm{d}V \tag{1.27}$$

where $s$ is the closed surface containing $V$ and $\mathbf{n}$ is the outward unit normal to $s$ at an element of area $\mathrm{d}s$. From this integral we can see that the Poynting vector

$$\mathbf{S} = \varepsilon_0 c^2 \mathbf{E}\times\mathbf{B} \tag{1.28}$$

must give the rate of flow of electromagnetic energy per unit area at each point on the surface $s$. Strictly speaking, our interpretation of $\mathbf{S}$ is tied up with a surface integral of the type in Eqn (1.27) but we can regard $\mathbf{S}$ as a vector giving the rate of flow of electromagnetic energy at a point in space.

Now that we have established the meaning of the Poynting vector $\mathbf{S}$ we should be able to identify the other term in Eqn (1.25). We said that each of the two terms in this equation gave the rate of change of energy per unit volume at a point in the electromagnetic field. This is the meaning of the time derivative of

$$\frac{\varepsilon_0}{2}(\mathbf{E}^2 + c^2\mathbf{B}^2)$$

The negative sign of this term in Eqn (1.25) simply shows that its rate of change is opposite in sense to $\nabla\cdot\mathbf{S}$, so allowing for the conservation of energy. Then the term

$$\frac{\varepsilon_0}{2}(\mathbf{E}^2 + c^2\mathbf{B}^2)$$

must describe the energy per unit volume at a point in the field, in other words, the 'energy density' $\mathscr{E}$.

So we have found that the electromagnetic energy density $\mathscr{E}$ at a point in the electromagnetic field is given by

$$\boxed{\mathscr{E} = \frac{\varepsilon_0}{2}(\mathbf{E}^2 + c^2\mathbf{B}^2)\ \mathrm{J\ m^{-3}}} \qquad (1.29)$$

and the rate of flow of electromagnetic energy, per unit area, at a point in the field, is given by the Poynting vector $\mathbf{S}$ where

$$\boxed{\mathbf{S} = \varepsilon_0 c^2 \mathbf{E} \times \mathbf{B}\ \mathrm{W\ m^{-2}}} \qquad (1.30)$$

## PROBLEMS

1.

Make an estimate of the dipole moment of the $H^+Cl^-$ (hydrogen chloride) molecule assuming that the hydrogen ion is fully ionized and that the separation of the hydrogen and chlorine nuclei is $\sim 1.3 \times 10^{-10}$ m. Note: The answer obtained is considerably higher than the experimental value because of the over simplified charge distribution assumed above. Further information is given by Smyth.

2.

Calculate the maximum value of the electric field at a distance of $100 \times 10^{-10}$ m from the $H^+Cl^-$ molecule (as described in Problem 1). Calculate the field at the same distance from a free electron.

3.

Show that the energy required to rotate an electric dipole of moment $\mathbf{p}$ in an electric field $\mathbf{E}$ is given by $-\mathbf{p}\cdot\mathbf{E}$, where we take zero energy to correspond to the position where the dipole is at right angles to the field.

4.

Estimate the maximum dipole–dipole interaction energy (from the relation given in Problem 3) of two $H^+Cl^-$ molecules separated by approximately $100 \times 10^{-10}$ m.

## FURTHER READING

See Appendix A for full details of books given here.

A more detailed discussion of the Poynting vector and energy flow is given by Feynman. See also, Reitz & Milford, Landau & Lifshitz, and Stratton. Both Marion and Holt give notes on the experimental background to Maxwell's equations.

Chapter 2

# MAXWELL'S EQUATIONS IN FREE SPACE

We wish to show that the Maxwell equations allow us to have waves of **E** and **B** in free space. In the process of doing this we find that the waves have very well defined properties. From our preliminary discussion in Section 1.2 of the Maxwell equations, it comes as no great surprise to find that the electric and magnetic waves are inextricably locked together so that we cannot have a time varying electric field without an associated magnetic field and vice versa. To establish the general fact that the Maxwell equations tell us that such electromagnetic waves can travel in free space, our approach is to search for a wave equation in **E** and then to seek a similar wave equation in **B**.

The first problem is to decide what we mean by free space. For our purposes all we need to say is that it is a region where $\rho = 0$ and $\mathbf{J} = 0$, that is, where there are no electric charges and no electric currents. Under these conditions, the general Maxwell equations given in Chapter 1 take on a particularly simple form:

$$\nabla \cdot \mathbf{E} = 0 \tag{2.1}$$

$$\nabla \times \mathbf{E} = -\partial \mathbf{B}/\partial t \tag{2.2}$$

$$\nabla \cdot \mathbf{B} = 0 \tag{2.3}$$

$$c^2 \nabla \times \mathbf{B} = \partial \mathbf{E}/\partial t \tag{2.4}$$

## 2.1 WAVES OF E

To explore the equations, we start by taking the curl of Eqn (2.2).

$$\nabla \times (\nabla \times \mathbf{E}) = \nabla \times (-\partial \mathbf{B}/\partial t) \tag{2.5}$$

Assuming that **B** is a well behaved function of space and time we can reverse the order of differentiation in the terms on the right hand side. Applying the mathematical identity for a vector triple product to the term on the left hand side gives

$$\nabla(\nabla \cdot \mathbf{E}) - \nabla^2 \mathbf{E} = -\frac{\partial}{\partial t}(\nabla \times \mathbf{B}) \tag{2.6}$$

Equations (2.1) and (2.4) state that

$$\nabla \cdot \mathbf{E} = 0$$
$$c^2 \nabla \times \mathbf{B} = \partial \mathbf{E}/\partial t$$

and so we obtain

$$-\nabla^2 \mathbf{E} = -\frac{\partial}{\partial t}\left(\frac{1}{c^2}\frac{\partial \mathbf{E}}{\partial t}\right)$$

that is

$$\boxed{\nabla^2 \mathbf{E} = \frac{1}{c^2}\frac{\partial^2 \mathbf{E}}{\partial t^2}} \tag{2.7}$$

This equation has the same form as the general three dimensional wave equation (Chapter 1) in $\psi$

$$\nabla^2 \psi = \frac{1}{v^2}\frac{\partial^2 \psi}{\partial t^2}$$

Each of the vector components in Eqn (2.7) corresponds to an equation of this type. We can draw the general conclusion that the Maxwell equations show that we can have waves of **E** in free space. The only restriction on such waves is that they must have a velocity of $\pm c$ in free space, that is, the waves of **E** always travel with the velocity of light in free space. Although Eqn (2.7) deals only with the **E** field, we know from Chapter 1 that there must be an accompanying magnetic wave since the Maxwell equations show that there can be no time variation in **E** without a corresponding variation in **B**.

## 2.2 WAVES OF B

A similar method can be used to show that there can be waves of **B** in free space. You should try to work this through for yourself using Section 2.1 as a guide. Just in case you run into trouble, the solution is given here.

In this case we start by taking the curl of Eqn (2.4)

$$c^2 \nabla \times (\nabla \times \mathbf{B}) = \nabla \times \partial \mathbf{E}/\partial t \tag{2.8}$$

Assuming that **E** is a well behaved function of space and time and making use of the identity for a vector triple product gives

$$c^2[\nabla(\nabla \cdot \mathbf{B}) - \nabla^2 \mathbf{B}] = \frac{\partial}{\partial t}(\nabla \times \mathbf{E}) \tag{2.9}$$

Equations (2.2) and (2.3) are

$$\nabla \cdot \mathbf{B} = 0$$
$$\nabla \times \mathbf{E} = -\partial \mathbf{B}/\partial t$$

and substitution into Eqn (2.9) gives

$$c^2[-\nabla^2\mathbf{B}] = \frac{\partial}{\partial t}\left(-\frac{\partial \mathbf{B}}{\partial t}\right)$$

that is

$$\boxed{\nabla^2\mathbf{B} = \frac{1}{c^2}\frac{\partial^2 \mathbf{B}}{\partial t^2}} \tag{2.10}$$

This is a three dimensional wave equation for the magnetic field **B**. We conclude that we can have waves of **B** in free space which travel with a velocity $c$, the velocity of light in free space. We have again obtained a wave equation in terms of one field only, the magnetic field, but we know that there is a corresponding **E** wave tied to the **B** wave.

## 2.3 PLANE ELECTROMAGNETIC WAVES IN FREE SPACE

We now consider a particular type of wave, a plane monochromatic wave, and find how **E** and **B** are related in an electromagnetic wave which is travelling in free space.

### 2.3.1 If We Specify the Variation in Time Then the Variation in Space is also Specified

Consider a wave which has a known dependence on time (a sinusoidal wave of angular frequency $\omega$) which is being propagated in the $z$-direction only. Such a wave is described by the equation

$$\mathbf{E}(\mathbf{r}, t) = \mathbf{E}_s(z)\exp(i\omega t) \tag{2.11}$$

where the electric field at any point in space and time $\mathbf{E}(\mathbf{r}, t)$ is described by the product of the vector $\mathbf{E}_s(z)$, which is real and a function of $z$ only, and $\exp(i\omega t)$, where $\omega$ is the angular frequency. The electric field is described by taking only the real part of the term $\exp(i\omega t)$, that is $\cos\omega t$, since $\mathbf{E}(\mathbf{r}, t)$ must be a real physical quantity—the electric field measured at some point $(\mathbf{r}, t)$. So we have now put into mathematical terms the idea that the electric wave is to travel in the $z$-direction only, with a sinusoidal variation in time determined by the angular frequency $\omega$. This is a plane 'monochromatic' wave, since the wave consists of one fixed frequency.

If we now substitute this function from Eqn (2.11) into our general three dimensional equation for electric waves

$$\nabla^2\mathbf{E} = \frac{1}{c^2}\frac{\partial^2 \mathbf{E}}{\partial t^2}$$

we obtain

$$\exp(i\omega t)\nabla^2\mathbf{E}_s = \frac{1}{c^2}\mathbf{E}_s(-\omega^2)\exp(i\omega t) \tag{2.12}$$

The term $\exp(i\omega t)$ is unaffected by $\nabla^2$ and $\mathbf{E}_s(z)$ is unaffected by the time derivative.

In the case of our plane wave there is no variation of the electric vector $\mathbf{E}_s(z)$ with the $x$ or $y$ coordinates and therefore

$$\nabla^2\mathbf{E}_s = \left(\frac{\partial^2}{\partial x^2}+\frac{\partial^2}{\partial y^2}+\frac{\partial^2}{\partial z^2}\right)\mathbf{E}_s$$

$$= \frac{\mathrm{d}^2\mathbf{E}_s}{\mathrm{d}z^2} \tag{2.13}$$

Consequently, Eqn (2.12) becomes

$$\frac{\mathrm{d}^2\mathbf{E}_s}{\mathrm{d}z^2} = -\frac{\omega^2}{c^2}\mathbf{E}_s \tag{2.14}$$

As a vector equation, this contains three ordinary differential equations (one for each vector component **i**, **j**, **k**) and each is of the form

$$\frac{\mathrm{d}^2 f}{\mathrm{d}z^2} = -\frac{\omega^2}{c^2} f \tag{2.15}$$

where $f$ is a function of $z$ only. This equation says that after two differentiations of $f$ with respect to $z$, the function $f$ has changed by only a constant factor of $(-\omega^2/c^2)$. The solution must then be an exponential function of $z$, say

$$f = C\exp(\beta z) \tag{2.16}$$

where $C$ and $\beta$ are constants. Substitution of this proposed solution into Eqn (2.15) yields the condition

$$\beta^2 = -\frac{\omega^2}{c^2}$$

or

$$\beta = \pm i\frac{\omega}{c}$$

Therefore $f$ can be written as

$$f = a\exp(i\omega z/c) + b\exp(-i\omega z/c)$$

where $a$ and $b$ are arbitrary constants. This can be more neatly expressed as

$$f = C\exp(\pm i\omega z/c) \tag{2.17}$$

with the understanding that the $\pm$ in the solution indicates that $C$ represents two possible arbitrary constants. Consequently, the three solutions to the three differential equations contained in Eqn (2.14) will *each* be of the general form $C\exp(\pm i\omega z/c)$ and can only differ from each other by (at most) the values of the arbitrary constants, $C$. The solution to the equation

$$\frac{d^2\mathbf{E}_s}{dz^2} = -\frac{\omega^2}{c^2}\mathbf{E}_s$$

will then be

$$\mathbf{E}_s = \mathbf{i}C_1\exp(\pm i\omega z/c)+\mathbf{j}C_2\exp(\pm i\omega z/c)+\mathbf{k}C_3\exp(\pm i\omega z/c)$$

or

$$\mathbf{E}_s = \mathbf{E}_0\exp(\pm i\omega z/c) \tag{2.19}$$

where $\mathbf{E}_0$ represents an arbitrary constant vector. The equation for the plane electric wave has become

$$\begin{aligned}\mathbf{E}(\mathbf{r}, t) &= \mathbf{E}_s(z)\exp(i\omega t)\\ &= \mathbf{E}_0\exp(\pm i\omega z/c)\exp(i\omega t)\end{aligned} \tag{2.20}$$

We can see that having specified the time variation, the variation in space has followed immediately. But what is the significance of the $\pm$ in the solution? From what we've already said about the arbitrary constant $C$ we can obviously write Eqn (2.20) as

$$\begin{aligned}\mathbf{E}(\mathbf{r}, t) = {} & \mathbf{E}_A\exp(i\omega z/c)\exp(i\omega t)\\ & +\mathbf{E}_B\exp(-i\omega z/c)\exp(i\omega t)\end{aligned}$$

where $\mathbf{E}_A$ and $\mathbf{E}_B$ are arbitrary constant vectors. Combining the exponential products gives

$$\mathbf{E}(\mathbf{r}, t) = \mathbf{E}_A\exp[(i\omega/c)(z+ct)]+\mathbf{E}_B\exp[(-i\omega/c)(z-ct)] \tag{2.21}$$

$$= \mathbf{E}_A f(z+ct)+\mathbf{E}_B g(z-ct) \tag{2.22}$$

From the general discussion of the description of waves in Chapter 1 it is obvious that the $\pm$ in the solution simply allows a choice (by means of the arbitrary constants) in whether the wave travels in the positive or negative $z$-direction. It is also clear that once we choose the arbitrary constant vector we have fixed the direction of the electric field in the wave. As the time varies, the sign of the sinusoidal term will then allow the electric field to swing from a 'positive' to a 'negative' maximum value along this fixed direction.

### 2.3.2 The Plane Wave must be a Transverse Wave

In free space, the first Maxwell equation states that

$$\nabla\cdot\mathbf{E} = 0$$

Written out fully this is

$$\frac{\partial E_x}{\partial x}+\frac{\partial E_y}{\partial y}+\frac{\partial E_z}{\partial z} = 0 \tag{2.23}$$

where we are taking

$$\mathbf{E} = \mathbf{i}E_x+\mathbf{j}E_y+\mathbf{k}E_z \tag{2.24}$$

For the plane electric wave we know that none of its components is a function of $x$ or $y$ and in this case the first Maxwell equation reduces to

$$\frac{\partial E_z}{\partial z} = 0 \tag{2.25}$$

We found that the plane wave is described by

$$\mathbf{E} = \mathbf{E}_0 \exp[(i\omega/c)(\pm z+ct)]$$

where we can write

$$\mathbf{E}_0 = \mathbf{i}E_{0X}+\mathbf{j}E_{0Y}+\mathbf{k}E_{0Z} \tag{2.26}$$

Substitution into the condition contained in Eqn (2.25) gives

$$(\pm i\omega/c)E_{0Z} \exp[(i\omega/c)(\pm z+ct)] = 0 \tag{2.27}$$

where we know that $E_{0Z}$ is a constant. The *only* way in which this can be zero for all $z$ and $t$ is if $E_{0Z}$ is zero. The first Maxwell equation *insists* that the plane electric wave has no component along the $z$-axis, that is, there is no component of the electric field in the direction of propagation of the wave. In other words, the plane electric wave is a 'transverse' wave.

It is important to build a mental picture of this wave. We cannot properly represent the wave in a diagram because the wave *fills* a three dimensional space and the page only allows us a two dimensional representation. We need to build up the full three dimensional picture in our mind. The wave is to travel forwards (say) in the $z$-direction with time. If we could freeze time and travel anywhere over any plane parallel to the $x$-$y$ plane, then we would find the same value of the electric field and, throughout this plane, the direction of the electric field would be the same. This situation is repeated in any other parallel plane, the only possible difference being in the magnitude of the electric field; the only change in direction which is possible for the electric field is that it can reverse. Now consider what will happen if we travel in the $z$-direction with time frozen; we then observe a sinusoidal variation in the electric field. The field always points in plus or minus the same direction, that is, parallel to the $x$-$y$ plane with no $z$-component. Now imagine yourself standing in *any* fixed position in space as our wave travels by. What do you experience in terms of the electric field? You observe a simple sinusoidal variation with time. If you had a friend standing somewhere else, the only possible difference in your records would be a difference in phase, that is, the maxima in the two records might not occur at corresponding times. This may appear to be a long winded explanation, but it is very important to be able to fully appreciate what we mean by such a plane transverse wave and to be able to relate the physical wave to its mathematical description.

### 2.3.3 The Associated B Wave

From the preliminary discussion of Maxwell's equations in Chapter 1, we know that if there is an **E** field which varies with time, then there must be an associated **B** field. In free space the two fields are tied together by the relations

$$\nabla \times \mathbf{E} = -\partial \mathbf{B}/\partial t \tag{2.29}$$

$$c^2 \nabla \times \mathbf{B} = \partial \mathbf{E}/\partial t \tag{2.30}$$

We can now find the **B** field associated with our plane electric wave:

$$\mathbf{E}(\mathbf{r}, t) = \mathbf{E}_0 \exp[(i\omega/c)(\pm z + ct)]$$

which, since there is no $z$-component, is

$$\mathbf{E}(\mathbf{r}, t) = \{\mathbf{i}E_{0X} + \mathbf{j}E_{0Y} + \mathbf{k}0\} \exp[(i\omega/c)(\pm z + ct)] \tag{2.31}$$

From this expression for the electric field we obtain

$$\nabla \times \mathbf{E} = \begin{vmatrix} \mathbf{i} & \mathbf{j} & \mathbf{k} \\ \dfrac{\partial}{\partial x} & \dfrac{\partial}{\partial y} & \dfrac{\partial}{\partial z} \\ [E_{0X}] & [E_{0Y}] & 0 \end{vmatrix} \tag{2.32}$$

where we are using $[E_{0X}]$ as shorthand for $E_{0X} \exp[(i\omega/c)(\pm z + ct)]$ and $[E_{0Y}]$ has a similar meaning. But,

$$\nabla \times \mathbf{E} = -\partial \mathbf{B}/\partial t$$

therefore

$$\begin{aligned} -\frac{\partial \mathbf{B}}{\partial t} = \mathbf{i}&\left\{\frac{\partial(0)}{\partial y} - \frac{\partial[E_{0Y}]}{\partial z}\right\} \\ -\mathbf{j}&\left\{\frac{\partial(0)}{\partial x} - \frac{\partial[E_{0X}]}{\partial z}\right\} \\ +\mathbf{k}&\left\{\frac{\partial[E_{0Y}]}{\partial x} - \frac{\partial[E_{0X}]}{\partial y}\right\} \end{aligned} \tag{2.33}$$

The **k** component must be zero since $[E_{0Y}]$ and $[E_{0X}]$ are not functions of $x$ or $y$. Therefore

$$\begin{aligned} \frac{\partial \mathbf{B}}{\partial t} = \; & \mathbf{i}(\pm i\omega/c)E_{0Y} \exp[(i\omega/c)(\pm z + ct)] \\ & -\mathbf{j}(\pm i\omega/c)E_{0X} \exp[(i\omega/c)(\pm z + ct)] \\ & +\mathbf{k}0 \end{aligned} \tag{2.34}$$

Partial integration with respect to time gives

$$\begin{aligned}\mathbf{B} = \mathbf{i}\{(\pm 1/c)E_{0Y}\exp[(i\omega/c)(\pm z+ct)]+f_1(x, y, z)\}\\ -\mathbf{j}\{(\pm i/c)E_{0X}\exp[(i\omega/c)(\pm z+ct)]+f_2(x, y, z)\}\\ +\mathbf{k}\{f_3(x, y, z)\}\end{aligned} \quad (2.35)$$

The 'constants of integration' are in this case $f_1(x, y, z)$, $f_2(x, y, z)$, $f_3(x, y, z)$ and can be set to zero since we are interested in 'waves', that is, in time varying quantities. So the magnetic field associated with our plane electric wave is

$$\begin{aligned}\mathbf{B} = \mathbf{i}\{(\pm 1/c)E_{0Y}\exp[(i\omega/c)(\pm z+ct)]\}\\ -\mathbf{j}\{(\pm 1/c)E_{0X}\exp[(i\omega/c)(\pm z+ct)]\}\\ +\mathbf{k}\{0\}\end{aligned} \quad (2.36)$$

which again is a transverse wave since there is no component of **B** in the direction of propagation, the $z$-direction. But consider the vector product of **k** and **E**:

$$\mathbf{k}\times\mathbf{E} = \begin{vmatrix} \mathbf{i} & \mathbf{j} & \mathbf{k} \\ 0 & 0 & 1 \\ [E_{0X}] & [E_{0Y}] & 0 \end{vmatrix}$$

$$= \mathbf{i}\{-[E_{0Y}]\}-\mathbf{j}\{-[E_{0X}]\}+\mathbf{k}\{0\}$$

that is

$$\mathbf{k}\times\mathbf{E} = (\mp c)\mathbf{B} \quad (2.37)$$

so **E** and **B** must be mutually perpendicular and both are perpendicular to the direction of propagation. In addition, since the magnitude of $\mathbf{k}\times\mathbf{E}$ is the same as the magnitude of **E**, the magnitudes of **E** and **B** are related by

$$\boxed{E = cB} \quad (2.38)$$

But compare this result with the expression we had in Chapter 1 for the Lorentz force

$$\mathbf{F} = q\mathbf{E}+q\mathbf{v}\times\mathbf{B}$$

We can see that the force experienced by an electric charge in the fields of an electromagnetic wave will be almost entirely due to the electric field unless the charge is moving with a velocity which is an appreciable fraction of the velocity of light.

We have now seen that Eqn (2.31) can represent the electric field in an electromagnetic wave travelling in free space, while Eqn (2.36) will represent the associated magnetic field. These equations fix the fields along the $\pm$ directions of two mutually perpendicular lines in space. There can then be no change in these directions as the wave travels through free space, in other words, the fields do not rotate and the direction of propagation does not change.

Since both the electric and magnetic fields are transverse to the direction of propagation, we call this type of wave a transverse electromagnetic wave or TEM wave.

The variation of the electric and magnetic field vectors is shown in Fig. 2.1 for a TEM wave as we travel in the $z$-direction (the direction of propagation) with time frozen. Remember that we would observe the same variation for any parallel line of travel and check that you can picture the full three dimensional situation in your mind.

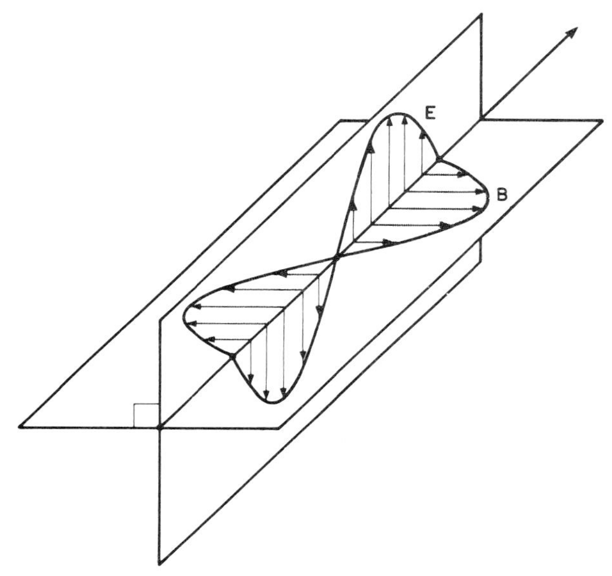


Figure 2.1

## EXERCISE 2.1

We have obtained these results by starting with a plane electric wave. See if you can obtain them starting from a plane magnetic wave. The working is essentially the same as that given above.

## EXERCISE 2.2

If we take the time average value of the Poynting vector and divide by the time average value of the energy density, the resulting velocity is known as the *energy velocity* $v_{en}$. By computing **S** and $\mathscr{E}$ from the real parts of the complex expressions for **E** and **B**, show that the energy velocity is $c$ for a simple plane electromagnetic wave in free space.

**Solution**

We can describe the electric field in a plane electromagnetic wave by the equation

$$\mathbf{E} = \mathbf{i}E_0 \exp[i\omega(t-z/c)]$$

where we are specifying that the wave is travelling in the $z$-direction with an electric field of amplitude $E_0$ along the $\pm x$-direction. Using the approach taken in Section 2.3.3 above, and the Maxwell equation

$$\nabla \times \mathbf{E} = -\partial \mathbf{B}/\partial t$$

we find the associated magnetic field as

$$\mathbf{B} = \mathbf{j}\left(\frac{1}{c}\right)E_0 \exp[i\omega(t-z/c)]$$

Taking the real parts of these complex expressions for $\mathbf{E}$ and $\mathbf{B}$ gives the Poynting vector $\mathbf{S}$ as

$$\begin{aligned}\mathbf{S} &= \varepsilon_0 c^2 \mathbf{E}\times\mathbf{B}\ (\mathrm{J\ s^{-1}\ m^{-2}})\\ &= \varepsilon_0 c^2 \mathbf{k}E_0 \cos[\omega(t-z/c)]\left(\frac{1}{c}\right)E_0\cos[\omega(t-z/c)]\end{aligned}$$

The time average is (see Appendix B)

$$\overline{\mathbf{S}} = \mathbf{k}\varepsilon_0 c E_0^2(\tfrac{1}{2})$$

The energy density $\mathscr{E}$ is given by

$$\mathscr{E} = \frac{\varepsilon_0}{2}(\mathbf{E}^2 + c^2\mathbf{B}^2)\ (\mathrm{J\ m^{-3}})$$

which, for this plane electromagnetic wave, is

$$\begin{aligned}\mathscr{E} &= \frac{\varepsilon_0}{2}\left\{E_0^2\cos^2[\omega(t-z/c)] + c^2\left(\frac{1}{c}\right)^2 E_0^2\cos^2[\omega(t-z/c)]\right\}\\ &= \varepsilon_0 E_0^2 \cos^2[\omega(t-z/c)]\end{aligned}$$

and the time average is then

$$\overline{\mathscr{E}} = \varepsilon_0 E_0^2(\tfrac{1}{2})$$

The energy velocity is then

$$\begin{aligned}v_{\mathrm{en}} &= \overline{\mathbf{S}}/\overline{\mathscr{E}} = \mathbf{k}\varepsilon_0 c E_0^2(\tfrac{1}{2})/\varepsilon_0 E_0^2(\tfrac{1}{2})\\ &= \mathbf{k}c\end{aligned}$$

The result shows the energy velocity to be in the direction of propagation and to have a magnitude $c$, the velocity of electromagnetic waves in free space.

## 2.4 SUMMARY OF RESULTS

Let's see what we've found out about electromagnetic waves in free space:

1. We found that the Maxwell equations show that we can have waves of **E** and waves of **B** in free space and that these waves must travel with a velocity $c$, the velocity of light in free space.

2. For plane monochromatic waves in free space we found that:

(i) the wave is transverse, that is, the direction of the fields is perpendicular to the direction of propagation.

This is called a transverse electromagnetic wave or TEM wave.

(ii) that there is no rotation of the wave as it travels, that is, once we fix the directions of the electric and magnetic fields these do not change as the wave travels. (We should note that by superposition of *two* waves which are 'out of phase' we can get rotation of the combined vectors as the combined wave travels.)

(iii) the magnitudes of the field vectors are related by $E = cB$ and this leads to the conclusion that the electric force is generally dominant in effect in an electromagnetic wave.

## PROBLEMS

1.
A commercially available laser is specified to have a beam size (cross sectional area) of $6 \times 25$ mm and to be capable of a maximum power of 1 MW in pulses of 10 ns duration. Knowing that a laser is a source of coherent radiation (the electromagnetic radiation is all of the same phase) and that the radiation is very nearly monochromatic, make an estimate of the maximum electric field in the beam. Use the expression for the Poynting vector and the free-space relation $E = cB$.

2.
The amount of radiant energy reaching the surface of the earth from the sun is approximately 1.2 kW $m^{-2}$. For the purpose of making the calculation, assume that the radiation is a monochromatic, linearly polarized plane wave striking the surface of the earth at normal incidence. On this basis, make an estimate of the amplitude of the electric field in the radiation.

## Chapter 3

# MAXWELL'S EQUATIONS IN MATERIALS

When we considered the Maxwell equations in free space, we started by defining the electric properties of that medium to be such that $\rho = 0$ and $\mathbf{J} = 0$. Now we need to develop some simple ideas about the electrical properties of a material medium. The general form of Maxwell's equations can be written in a more convenient way for a material if we introduce the polarization vector **P**. This leads to the definition of the relative permittivity $\varepsilon$ for a material and we can show that $\varepsilon$ is simply related to the refractive index of the material, $n$.

## 3.1 A MODEL FOR THE INDIVIDUAL MOLECULE

Although we will talk in terms of the more general and useful case of a molecule, all that we have to say could equally well be applied to an atom like helium, for example. We take the molecule to consist of both positive and negative charge. There are many types of material in which the molecules show a distinct and permanent separation of charge, for example, hydrogen chloride $H^+Cl^-$. In our simple model we will consider molecules which do *not* have such a charge separation, for example, methane $CH_4$. We regard such molecules as being electrically neutral so that the positive charge is balanced by negative charge and the molecule appears to have zero net charge. However, the application of an electric field will cause a separation of the charges and so produce an electric dipole within the molecule. It is this charge separation that concerns us.

Assume that in the presence of an electric field, a charge $q$ is moved (within the molecule) through a distance $x$ from its equilibrium position. If the displaced charge has a mass $m$ and experiences a restoring force proportional to the displacement, the force equation for the charge can be written as

$$qE = m\left(\frac{\mathrm{d}^2x}{\mathrm{d}t^2} + \gamma\frac{\mathrm{d}x}{dt} + \omega_0^2 x\right) \tag{3.1}$$

where $E$ is the magnitude of the electric field experienced by the charge and the displacement $x$ is in the direction of **E**. This force equation also allows for a velocity dependent damping force given by $m\gamma(\mathrm{d}x/\mathrm{d}t)$; the other terms are the restoring force $m\omega_0^2 x$ and the acceleration term $m(\mathrm{d}^2x/\mathrm{d}t^2)$.

To go any further we must specify how **E** behaves as a function of time. Consider a sinusoidally varying field (like the plane wave in Chapter 2) and write the time dependence of **E** as

$$E = E_0 \exp(i\omega t) \tag{3.2}$$

where $E_0$ is the magnitude of the constant vector $\mathbf{E}_0$. In this case we do not have to concern ourselves with the $z$-dependence of **E** since the molecule can be taken to be in a fixed position. We can assume that in the steady state the charge oscillates with the same frequency as the field driving it and so we can propose that the displacement is given by

$$x = x_0 \exp(i\omega t) \tag{3.3}$$

where $x_0$ is a constant and where, as usual, the physical quantity (the displacement) is found by taking only the real part of this complex quantity. Substitution of Eqn (3.3) into the force equation (Eqn (3.1)) gives

$$qE = m[-\omega^2 x + \gamma(i\omega x) + \omega_0^2 x] \tag{3.4}$$

and therefore

$$x = \frac{qE/m}{-\omega^2 + i\omega\gamma + \omega_0^2} \tag{3.5}$$

The imaginary term associated with $\gamma$ indicates that the damping in our model causes the displacement to be out of phase with the driving force of the electric field. It is a simple matter to obtain the real part of this expression for $x$, but the information about the relative phase of $x$ and **E** is very neatly stored in the complex form and it is convenient to leave our expression for $x$ as it is.

The action of the oscillating electric field is to produce an oscillating electric dipole within the molecule. We had a brief discussion of electric dipoles in Exercise 1.2 where the term 'dipole moment' was defined. We can similarly define the molecular electric dipole moment, $\mathbf{p}_m$, so that

$$p_m = qx \tag{3.6}$$

and

$$\boxed{\mathbf{p}_m = \frac{q^2\mathbf{E}(t)/m}{-\omega^2 + i\omega\gamma + \omega_0^2}} \tag{3.7}$$

The molecular dipole moment is a function of time and is in the direction of the field **E**. This expression can be tidied up a little by replacing some of the constants by a term $\alpha$ called the molecular polarizability, where

$$\alpha = \frac{q^2/m\varepsilon_0}{-\omega^2 + i\gamma\omega + \omega_0^2} \tag{3.8}$$

Then

$$\boxed{\mathbf{p}_m = \varepsilon_0 \alpha \mathbf{E}} \tag{3.9}$$

where $\alpha$ is a function of the nature of the molecule and of the angular frequency $\omega$ of the applied field $\mathbf{E}$.

All of this relates to an individual molecule and so forms the basis of our 'microscopic' description of a material.

## 3.2 THE POLARIZATION VECTOR P

The discussion of our general molecular model for a material suggests that there may be two distinct sets of conditions for which we may wish to describe the behaviour of charge in a material. We have been discussing molecular charges in terms of their displacement from the equilibrium position in the molecule. Although such charges are capable of movement, they are restricted to the confines of the molecule—we are not concerned with the very high field strengths that could pull a charge free and lead to 'electrical breakdown' within the material. These then are charges which belong to the molecules of the material. Other charges can be introduced into the material—free ions or free electrons, for example. Such charges can be described as 'free charges' since they are not restricted to move within the limits of a molecule. On this basis we can take account of the two types of charge by writing

$$\rho = \rho_f + \rho_m \tag{3.10}$$

where we are dividing the total charge density $\rho(\mathbf{r}, t)$ into $\rho_f(\mathbf{r}, t)$ the 'free' charge density and $\rho_m(\mathbf{r}, t)$ the molecular charge density, sometimes known as the 'bound' charge density.

The total current density $\mathbf{J}$ can be similarly divided as

$$\mathbf{J} = \mathbf{J}_f + \mathbf{J}_m \tag{3.11}$$

where $\mathbf{J}_f(\mathbf{r}, t)$ is the free current density and $\mathbf{J}_m(\mathbf{r}, t)$ is the molecular or 'bound' current density. We now introduce a vector $\mathbf{P}$ to describe the movement of the molecular charge. We *define* $\mathbf{P}(\mathbf{r}, t)$ as a vector giving the direction of flow of *molecular charge* at any point $(\mathbf{r}, t)$; the magnitude of $\mathbf{P}$ is defined to be equal to the quantity of molecular charge, per unit area, which has passed the point $(\mathbf{r}, t)$ since the medium had a neutral distribution of molecular charge. So $\mathbf{P}$ is monitoring the flow of charge at each point. Its value at any time is determined by the net flow of charge past that point since the material had a neutral charge distribution. We can see that the current density $\mathbf{J}_m$ associated with this movement of molecular charge, is given by

$$\mathbf{J}_m = \frac{\partial \mathbf{P}}{\partial t} \tag{3.12}$$

Now consider a volume $V$ of the material, bounded by a surface $s$. If the material becomes polarized we can generally assume that the flow of charge into the volume $V$ will equal any flow of charge out of $V$. The total flow can be measured by monitoring the flow across each surface element $\mathrm{d}s$ on $s$ as shown in Fig. 3.1.

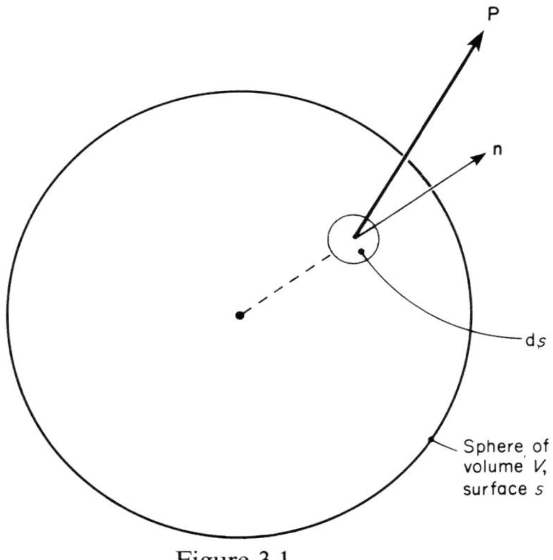


Figure 3.1

Calling $\sigma$ the amount of charge which has crossed $\mathrm{d}s$ since neutrality, we can write

$$\sigma = \mathbf{P}\cdot\mathbf{n}\,\mathrm{d}s$$

where $\mathbf{n}$ is the outward unit normal to $\mathrm{d}s$. We are regarding the outward flow of positive charge as a positive quantity since in that case $\mathbf{P}\cdot\mathbf{n}$ will be positive and will have the magnitude of $\mathbf{P}$ if $\mathbf{P}$ and $\mathbf{n}$ are in the same direction. The total amount of positive charge which has flowed out of $V$ is given by

$$\Delta Q = \int_s \mathbf{P}\cdot\mathbf{n}\,\mathrm{d}s \tag{3.14}$$

If this is the amount of charge which has left the volume since neutrality then a similar amount of negative charge must now be left in $V$. This residual charge is given by the charge density within $V$, say $\rho_m(\mathbf{r}, t)$. Its total value is given by

$$-\Delta Q = \int_V \rho_m(\mathbf{r}, t)\,\mathrm{d}V \tag{3.15}$$

where $\mathrm{d}V$ is a volume element of $V$. We now have two expressions for $\Delta Q$ and they must be equal,

$$\int_s \mathbf{P}\cdot\mathbf{n}\,\mathrm{d}s = -\int_V \rho_m(\mathbf{r}, t)\,\mathrm{d}V \tag{3.16}$$

Applying Gauss' Integral Theorem to the left hand side gives

$$\int_V \nabla\cdot\mathbf{P}\,\mathrm{d}V = -\int_V \rho_m\,\mathrm{d}V \tag{3.17}$$

This relation is general for *any* volume $V$ and therefore we must have

$$\boxed{\nabla\cdot\mathbf{P} = -\rho_m} \tag{3.18}$$

Then the divergence of the polarization vector **P** is related to the charge density $\rho_m$ and the time derivative of **P** gives the current density $\mathbf{J}_m$.

All of this is quite general and is independent of the details of the material's structure. This means that we can now write the Maxwell equations in a form which is particularly suitable for a material medium.

The general form of the first Maxwell equation is

$$\nabla\cdot\mathbf{E} = \rho/\varepsilon_0 \tag{3.19}$$

which can now be written as

$$\begin{aligned}\nabla\cdot\mathbf{E} &= (\rho_f+\rho_m)/\varepsilon_0\\ &= [\rho_f+(-\nabla\cdot\mathbf{P})]/\varepsilon_0\end{aligned}$$

that is

$$\nabla\cdot(\mathbf{E}+\mathbf{P}/\varepsilon_0) = \rho_f/\varepsilon_0 \tag{3.20}$$

The general form of the fourth Maxwell equation is

$$c^2\nabla\times\mathbf{B} = \mathbf{J}/\varepsilon_0+\partial\mathbf{E}/\partial t$$

and this can now be written as

$$\begin{aligned}c^2\nabla\times\mathbf{B} &= \mathbf{J}_f/\varepsilon_0+\mathbf{J}_m/\varepsilon_0+\partial\mathbf{E}/\partial t\\ &= \mathbf{J}_f/\varepsilon_0+\left(\frac{1}{\varepsilon_0}\right)\frac{\partial\mathbf{P}}{\partial t}+\frac{\partial\mathbf{E}}{\partial t}\end{aligned}$$

that is

$$c^2\nabla\times\mathbf{B} = \mathbf{J}_f/\varepsilon_0+\frac{\partial}{\partial t}(\mathbf{E}+\mathbf{P}/\varepsilon_0) \tag{3.22}$$

Then, in a material medium we can write the four Maxwell equations as:

$$\boxed{\begin{aligned}\nabla\cdot(\mathbf{E}+\mathbf{P}/\varepsilon_0) &= \rho_f/\varepsilon_0\\ \nabla\times\mathbf{E} &= -\partial\mathbf{B}/\partial t\\ \nabla\cdot\mathbf{B} &= 0\\ c^2\nabla\times\mathbf{B} &= \mathbf{J}_f/\varepsilon_0+\frac{\partial}{\partial t}(\mathbf{E}+\mathbf{P}/\varepsilon_0)\end{aligned}} \tag{3.23}$$

This new vector **P**, like the other terms in the Maxwell equations, has a value at each point (**r**, $t$). Within a material we may expect that **P** will vary rapidly from point to point as indeed may the fields **E** and **B**. In general any measurement which we can make on the material will involve a large number of molecules and a volume which is large in comparison to the size of a molecule. For example, if the measurement is to determine the flow of charge within the material when it becomes polarized, we can expect that the measurement will determine the *average* value of **P**.

## 3.3 OUR MOLECULAR MODEL FOR A MATERIAL AND P

In addition to the relationship between **P** and the terms $\rho_m$ and $\mathbf{J}_m$ we would also like to relate **P** to the molecular dipole moment $\mathbf{p}_m$. Suppose the material contains $N$ molecules per unit volume (1 m$^3$). Consider a plane surface of area $A$ in the material, perpendicular to the direction of polarization (the $x$-direction). If each molecular charge $q$ moves a distance $x$ on polarization of the medium by a field **E**, then the total charge which crosses the surface $A$ is (on average) $qNxA$. The situation is illustrated in Fig. 3.2.

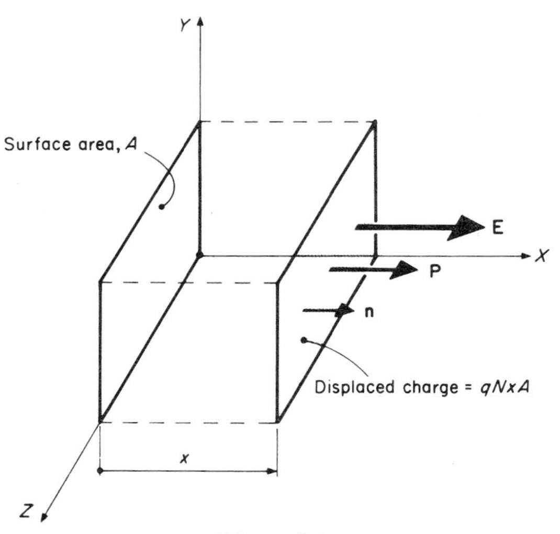


Figure 3.2

But if we integrate **P** across the surface, then this should also give the value for the charge since we defined **P** to measure just such a movement of charge, that is,

$$qNxA = \int_A P \, \mathrm{d}A \tag{3.24}$$

where $\mathrm{d}A$ is an element of $A$. We can also write

$$\int_A P\,\mathrm{d}A = P_{av}A \tag{3.25}$$

where $P_{av}$ is the average value of $P$ over the area $A$. We must deal with the average value because $\mathbf{P}(\mathbf{r}, t)$ is a vector with a value at each point and if the material is composed of discrete molecules which are well separated in space (for example, a low density gas) then the value of $\mathbf{P}$ will vary from point to point on the surface $A$. Then

$$qNxA = P_{av}A \tag{3.26}$$

that is

$$Np_m = P_{av}$$

or

$$\mathbf{P}_{av} = N\mathbf{p}_m \tag{3.27}$$

where $\mathbf{P}_{av}$ and $\mathbf{p}_m$ are in the same direction.

In Section 3.1 we obtained the result

$$\mathbf{p}_m = \varepsilon_0\alpha\mathbf{E} \tag{3.28}$$

and so we can now write

$$\boxed{\mathbf{P}_{av} = \varepsilon_0 N\alpha\mathbf{E}} \tag{3.29}$$

Equation (3.28) was obtained for an individual molecule on the basis that the field producing the molecular polarization is $\mathbf{E}$. This is reasonable if the density of the material is sufficiently low to ensure that the polarization of one molecule does not influence the field which is then experienced by its neighbours. Apart from occasional collisions between molecules, this is generally the situation in a low density gas. We now consider the case of a dense material where the molecules are sufficiently close to influence each other.

## 3.4 THE ELECTRIC FIELD WITHIN A DENSE MATERIAL

We can expect that once a material becomes polarized, the resulting charge distribution will mean that the calculation of the electric field at a point within the material may become a difficult task. This is particularly the case in materials where the molecules contain permanent dipoles. In principle we could attempt the calculation using the most general form of Maxwell's equations and assuming that we know the positions of *all* the charges within the material. An alternative and more feasible method is possible if we continue to restrict ourselves to relatively simple materials where the molecules do not have permanent dipoles.

Consider a material which consists of neutral molecules in the gas or liquid state. Because the material is a fluid its properties can generally be taken as isotropic and if we apply an electric field we can expect to produce polarization of the material as a result of the charge separation occurring in the individual molecules. The polarization will be in the direction of the applied field. In the electrostatic case where the material is contained between the plates of a parallel plate capacitor, the field at a point within the material will be caused by, (i) the charge distribution at the external surfaces of the material consisting of the charge on the plates and the surface polarization of the material and, (ii) an additional contribution from the molecular dipoles surrounding the point of interest. Contribution (i) is simply determined from the *net* surface charge per unit area, $\sigma$ C m$^{-2}$, as

$$E_{(i)} = \sigma/\varepsilon_0 \tag{3.30}$$

This result was obtained in Exercise 1.3. The situation is illustrated in Figs 3.3(a) and 3.3(b) for a parallel plate capacitor charged from a constant voltage supply. Introducing the dielectric causes additional charge to flow from the ideal

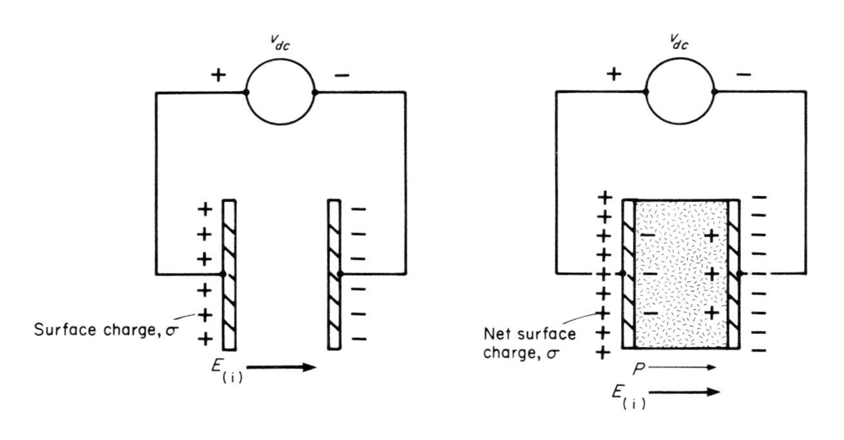


(a) Unfilled parallel plate capacitor (b) Filled parallel plate capacitor

Figure 3.3

voltage source to balance up the surface polarization charge of the dielectric and so maintain the voltage difference. There is still a *net* surface charge $\sigma$ at the capacitor plates and the field due to *this* charge distribution is still given by Eqn (3.30). To evaluate contribution (ii), we follow the method of Lorentz and draw a sphere around the point of interest as illustrated in Fig. 3.4(a). Inside the sphere we consider the material at the detailed level of individual molecules. Outside the sphere we regard the material as being effectively neutral. This is reasonable since (as we've seen in Exercise 1.2) the electric field due to the individual dipoles falls off rapidly as $r^{-3}$ and will be negligible at a sufficiently large distance. The (imaginary) spherical surface of the 'continuous' material will appear to have a surface charge determined by **P**. To find the field $\mathbf{E}_s$ due to the spherical surface charge we take a circular strip around the direction of the field, the $x$-direction, say, as shown in Fig. 3.4(b). If the width of the strip

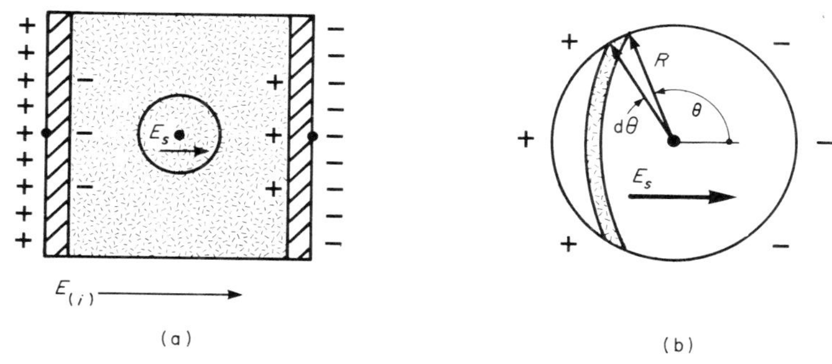


Figure 3.4

is $R\,d\theta$ then it contains a surface charge of $(P\cos\theta)2\pi(R\sin\theta)R\,d\theta$. This charge produces a field at the centre of the sphere with an $x$-component (only) equal to

$$\frac{[(P\cos\theta)2\pi(R\sin\theta)R\,d\theta]\cos\theta}{(4\pi\varepsilon_0)R^2}$$

The total field at the centre of the sphere, due to its surface charge, is found by integrating over the entire surface to give

$$\int_{\theta=0}^{\theta=\pi}\left\{-\frac{P\cos^2\theta\,d(\cos\theta)}{2\varepsilon_0}\right\} = \left[-\frac{P_{av}\cos^3\theta}{6\varepsilon_0}\right]_{\theta=0}^{\theta=\pi}$$

$$= P_{av}/3\varepsilon_0 \qquad (3.31)$$

The integration naturally gives the average value of the polarization.

We now need the field due to the individual molecular dipoles within the sphere. Using the expression we obtained in Exercise 1.2 for the field due to an electric dipole we can write the field for each molecular dipole $\mathbf{p}_m$ as

$$\mathbf{E} = \frac{1}{4\pi\varepsilon_o}\left[\frac{3(\mathbf{p}_m\cdot\mathbf{r})\mathbf{r}}{r^5} - \frac{\mathbf{p}_m}{r^3}\right]$$

where in this case it is convenient to take the position vector **r** as drawn from the field point (the centre of the sphere) to the position of the dipole inside the sphere. The induced molecular dipoles are all in the direction of the electric field (the

$x$-direction) and we can write their total field at the centre of the sphere as

$$\frac{1}{4\pi\varepsilon_0}\sum\left\{\frac{3(p_{mx}x)(\mathbf{i}x+\mathbf{j}y+\mathbf{k}z)}{r^5}-\mathbf{i}\left(\frac{p_{mx}}{r^3}\right)\right\} \quad (3.32)$$

$$=\frac{1}{4\pi\varepsilon_0}\left(\frac{p_{mx}}{r^3}\right)\mathbf{i}\sum\left(\frac{3x^2}{r^2}-1\right)+\frac{3}{4\pi\varepsilon_0}\left(\frac{p_{mx}}{r^3}\right)\sum\left(\mathbf{j}\frac{xy}{r^2}+\mathbf{k}\frac{xz}{r^2}\right) \quad (3.33)$$

Since $r^2 = x^2+y^2+z^2$ and since we have a large number of molecules within the sphere, the average value of $x^2$ is $r^2/3$ and the average values of $xy$ and $xz$ are zero. Consequently the value of the field contribution given in Eqn (3.33) is zero.

We are left with only two contributions to the field at the centre of the sphere arising from (i) the charge at the external (plate) surface of the material and (ii) the polarization charge on the inside of the spherical surface. The local field is then given by

$$\mathbf{E}_{local} = \mathbf{E}_{(i)}+\mathbf{E}_{(ii)}$$

which we find to be

$$\boxed{\mathbf{E}_{local} = \mathbf{E}_{(i)}+\mathbf{P}_{av}/3\varepsilon_0} \quad (3.34)$$

where in the present case $E_{(i)} = \sigma/\varepsilon_0$ and $\sigma$ is the net surface charge per unit area at the surface of the material. This expression is known as the Lorentz Local Field. It shows that local effects can increase the field by $\mathbf{P}_{av}/3\varepsilon_0$.

## 3.5 REFRACTIVE INDEX AND RELATIVE PERMITTIVITY

We *define* the refractive index $n$ of a material as

$$n = c/v \quad (3.35)$$

where $c$ is the velocity of an electromagnetic wave in free space and $v$ is the velocity of the wave in the material of refractive index $n$. This quantity will become more meaningful in the next and later chapters but we now go on to define another quantity for the material—the relative permittivity $\varepsilon$. This is defined by

$$\varepsilon = \frac{E+P/\varepsilon_0}{E} \quad (3.36)$$

We can easily show (see Exercise 3.1 below) that in the electrostatic case, $\varepsilon$ corresponds to the factor by which the capacitance is increased in a parallel plate capacitor when a material is introduced to fill the space between the plates. At the moment we will concentrate on establishing the general relationship between $n$ and $\varepsilon$. Setting the free charge density $\rho_f$ to zero and setting the free current density $\mathbf{J}_f$ to zero reduces the Maxwell equations (Eqns (3.23)) for a material to

$$\nabla\cdot(\mathbf{E}+\mathbf{P}/\varepsilon_0) = 0 \tag{3.37}$$

$$\nabla\times\mathbf{E} = -\partial\mathbf{B}/\partial t \tag{3.38}$$

$$\nabla\cdot\mathbf{B} = 0 \tag{3.39}$$

$$c^2\nabla\times\mathbf{B} = \frac{\partial}{\partial t}(\mathbf{E}+\mathbf{P}/\varepsilon_0) \tag{3.40}$$

From the definition of $\varepsilon$ we can write the last equation as

$$c^2\nabla\times\mathbf{B} = \varepsilon\,\partial\mathbf{E}/\partial t$$

and taking the curl gives

$$c^2\nabla\times(\nabla\times\mathbf{B}) = \varepsilon\frac{\partial}{\partial t}(\nabla\times\mathbf{E})$$

where as usual it is assumed that $\mathbf{E}$ is a well behaved function of space and time so allowing us to reverse the order of taking the space and time derivatives. Using the mathematical identity for the vector triple product and Eqn (3.38) gives

$$c^2[\nabla(\nabla\cdot\mathbf{B})-\nabla^2\mathbf{B}] = \varepsilon\frac{\partial}{\partial t}\left(-\frac{\partial\mathbf{B}}{\partial t}\right)$$

which from Eqn (3.39) is

$$\nabla^2\mathbf{B} = \frac{\varepsilon}{c^2}\frac{\partial^2\mathbf{B}}{\partial t^2} \tag{3.41}$$

This is the three dimensional wave equation in $\mathbf{B}$ with the velocity of the wave, $v$, given by

$$v^2 = 1/(\varepsilon/c^2)$$

But the definition of the refractive index $n$ is

$$n = c/v$$

therefore

$$v^2 = c^2/n^2$$

Hence

$$\boxed{n^2 = \varepsilon} \tag{3.42}$$

## EXERCISE 3.1

Show that when the space separating the plates of a parallel plate capacitor is filled with a material of relative permittivity $\varepsilon$, then the value of its electrostatic capacitance increases by a factor of $\varepsilon$.

**Solution**

The capacitance $C$ is defined by the relation

$$C = Q/V$$

where $Q$ is the charge on the positive plate and $V$ is the 'potential difference' between the plates. The potential of a point is defined as the work done per unit positive charge in bringing a charge from infinity to the point. For a parallel plate capacitor with a uniform field $\mathbf{E}$ the potential difference is obviously $E \cdot d$ where $d$ is the separation of the plates. None of this really need concern us too much since we can take it that the capacitor is supplied from a constant voltage

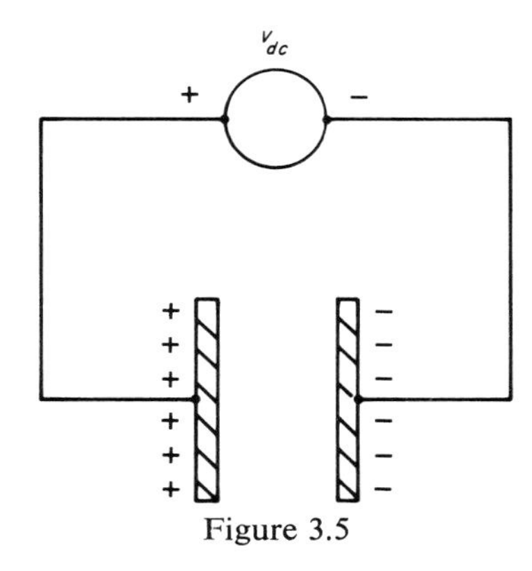


Figure 3.5

source (see Fig. 3.5) during the 'measurement'. Then the factor by which the capacitance increases is

$$\frac{C}{C_0} = \frac{Q}{Q_0}$$

where $C_0$ and $Q_0$ are the empty capacitance and charge values, and $C$ and $Q$ are the corresponding values when the capacitor is filled. If the charge per unit area on the positive capacitor plate is $\sigma_0$ and $\sigma$ with the capacitor empty and filled (see Fig. 3.6) we can write

$$\frac{C}{C_0} = \frac{\sigma}{\sigma_0}$$

The problem is now reduced to deciding by how much does $\sigma$ increase when the capacitor is filled. The effect of polarization is to attempt to *reduce* the *net* charge by $P$; but if the potential difference is maintained then the electric field is maintained. The electric field is given by the net surface charge divided by $\varepsilon_0$ (as we've seen in Exercise 1.3). To maintain the *net* surface charge there must be an additional flow of charge to the plates to *increase* $\sigma_0$ by $P$, that is

$$\sigma = \sigma_0 + P$$

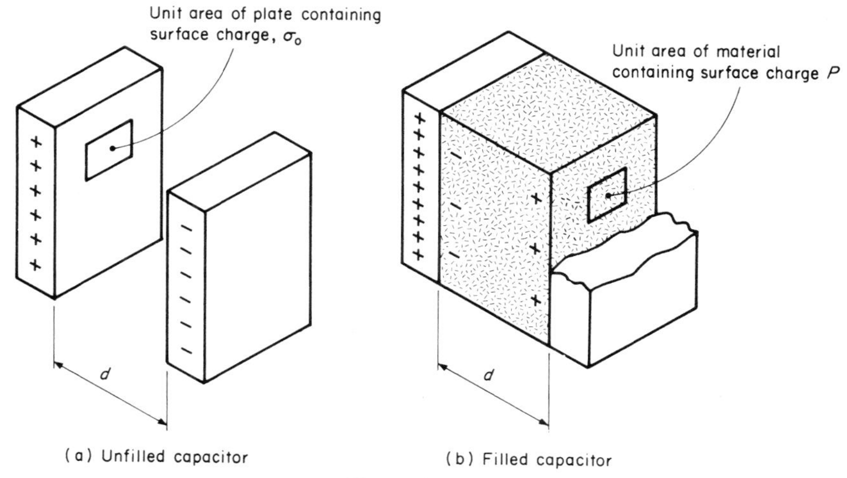


Figure 3.6

Then

$$\frac{C}{C_0} = \frac{\sigma}{\sigma_0}$$

$$= \frac{\sigma_0 + P}{\sigma_0}$$

$$= \frac{\sigma_0/\varepsilon_0 + P/\varepsilon_0}{\sigma_0/\varepsilon_0}$$

$$= \frac{E_0 + P/\varepsilon_0}{E_0}$$

which from the definition of $\varepsilon$ is

$$\frac{C}{C_0} = \varepsilon$$

## 3.6 SUMMARY OF RESULTS

1. The induced molecular dipole moment is given by our simple model for a dielectric material as

$$\mathbf{p}_m = \frac{q^3/m}{(\omega_0^2 - \omega^2) + i\omega\gamma}\,\mathbf{E}(t)$$

2. The Maxwell equations can be written in terms of the polarization vector **P** as

$$\nabla\cdot(\mathbf{E}+\mathbf{P}/\varepsilon_0) = \rho_f/\varepsilon_0, \qquad \nabla\times\mathbf{E} = -\partial\mathbf{B}/\partial t$$

$$\nabla\cdot\mathbf{B} = 0, \qquad c^2\nabla\times\mathbf{B} = \mathbf{J}_f/\varepsilon_0+\frac{\partial}{\partial t}(\mathbf{E}+\mathbf{P}/\varepsilon_0)$$

3. In terms of our simple model the average polarization of a material is given by

$$\mathbf{P}_{av} = \varepsilon_0 N\alpha\mathbf{E} = N\mathbf{p}_m$$

4. The electric field in a dense material is given by the Lorentz expression

$$\mathbf{E}_{local} = \mathbf{E}+\mathbf{P}_{av}/3\varepsilon_0$$

5. The refractive index $n$ of a material is defined by

$$n = c/v$$

where $c$ is the velocity of an electromagnetic wave in free space and $v$ is its velocity in the material.

6. The relative permittivity $\varepsilon$ of a material is defined as

$$\varepsilon = \frac{E+P/\varepsilon_0}{E}$$

7. The relative permittivity and the refractive index are related by

$$\varepsilon = n^2$$

Chapter 4

# ELECTROMAGNETIC WAVES IN NON-CONDUCTING MATERIALS

In the previous chapter we found a convenient way of writing the Maxwell equations when they were applied to a material. We also developed a very simple model to describe the electrical behaviour of a material. Now we try to put these results together to find the conditions under which the Maxwell equations show that electromagnetic waves can travel through a material. The method is essentially the same as the one used to show that electromagnetic waves can propagate through free space, but in this case we obtain some interesting results for the material.

## 4.1 ELECTROMAGNETIC WAVES IN A LOW DENSITY GAS

We start by saying that there are no free charges and no free currents, that is, $\rho_f = 0$ and $\mathbf{J}_f = 0$. Then, from Chapter 3, the Maxwell equations can be written as

$$\nabla \cdot (\mathbf{E} + \mathbf{P}/\varepsilon_0) = 0 \tag{4.1}$$

$$\nabla \times \mathbf{E} = -\partial \mathbf{B}/\partial t \tag{4.2}$$

$$\nabla \cdot \mathbf{B} = 0 \tag{4.3}$$

$$c^2 \nabla \times \mathbf{B} = \frac{\partial}{\partial t}(\mathbf{E} + \mathbf{P}/\varepsilon_0) \tag{4.4}$$

The approach used in Chapter 2 can again be applied to look for a wave equation in $\mathbf{E}$. Taking the curl of Eqn (4.2) gives

$$\nabla \times (\nabla \times \mathbf{E}) = -\frac{\partial}{\partial t}(\nabla \times \mathbf{B}) \tag{4.5}$$

where, assuming that $\mathbf{B}$ is a well behaved function of space and time, the order of differentiation can be reversed. Using the mathematical identity for the vector triple product and substituting for $\nabla \times \mathbf{B}$ from Eqn (4.4), gives

$$\nabla(\nabla \cdot \mathbf{E}) - \nabla^2 \mathbf{E} = -\frac{1}{c^2}\frac{\partial^2 \mathbf{E}}{\partial t^2} - \frac{1}{c^2 \varepsilon_0}\frac{\partial^2 \mathbf{P}}{\partial t^2} \tag{4.6}$$

The divergence of **E** is given by Eqn (4.1) in terms of **P** and so we obtain

$$-\frac{1}{\varepsilon_0}\nabla(\nabla\cdot\mathbf{P})-\nabla^2\mathbf{E} = -\frac{1}{c^2}\frac{\partial^2\mathbf{E}}{\partial t^2}-\frac{1}{c^2\varepsilon_0}\frac{\partial^2\mathbf{P}}{\partial t^2}$$

or

$$\nabla^2\mathbf{E} = \frac{1}{c^2}\frac{\partial^2\mathbf{E}}{\partial t^2}+\frac{1}{c^2\varepsilon_0}\frac{\partial^2\mathbf{P}}{\partial t^2}-\frac{1}{\varepsilon_0}\nabla(\nabla\cdot\mathbf{P}) \tag{4.7}$$

At this stage we seem to be some way short of obtaining the three dimensional wave equation in **E**, that is, an equation of the type

$$\nabla^2\mathbf{E} = \frac{1}{v^2}\frac{\partial^2\mathbf{E}}{\partial t^2} \tag{4.8}$$

It is clear that the relationship between **P** and **E** will decide if Eqn (4.7) will reduce to this form. The problem appears to be a little more complicated than we may have anticipated! To simplify matters we can set up the equation for a monochromatic plane polarized wave and see if such a wave satisfies the general Eqn (4.7). The electric field in our plane wave is described by

$$\mathbf{E} = \mathbf{i}E_x = \mathbf{i}E_0\exp[i(\omega t-kz)] \tag{4.9}$$

where the electric vector is in the $\pm x$ direction and the velocity of the wave is $v = \omega/k$. The wave travels in the $z$-direction and $k$ is known as the propagation constant.

Since the refractive index $n$ of a material has been defined as

$$n = c/v$$

the equation for the electric field can be written as

$$\begin{aligned}\mathbf{E} &= \mathbf{i}E_x\\ &= \mathbf{i}E_0\exp[i\omega(t-zn/c)]\end{aligned} \tag{4.10}$$

Our simple model shows that within a material, the displacement of the charge is in the direction of the applied electric field and so the polarization **P** will also be in the $\pm x$-direction and will vary with an angular frequency $\omega$. Then in this case,

$$\begin{aligned}\nabla\cdot\mathbf{P} &= \frac{\partial P_x}{\partial x}+\frac{\partial P_y}{\partial y}+\frac{\partial P_z}{\partial z}\\ &= \frac{\partial P_x}{\partial x}\\ &= 0\end{aligned} \tag{4.11}$$

because **P** has (like **E**) an $x$-component only and because the $x$-component (like **E**) depends only on $z$ and $t$. The other term in **P** which appears in Eqn (4.7) is $\partial^2\mathbf{P}/\partial t^2$. In the case of our plane wave

$$\begin{aligned}\frac{\partial^2\mathbf{P}}{\partial t^2} &= \mathbf{i}\,\frac{\partial^2\mathbf{P}_x}{\partial t^2}+\mathbf{j}\,\frac{\partial^2\mathbf{P}_y}{\partial t^2}+\mathbf{k}\,\frac{\partial^2\mathbf{P}_z}{\partial t^2}\\ &= \mathbf{i}\,\frac{\partial^2\mathbf{P}_x}{\partial t^2}\\ &= \mathbf{i}(-\omega^2\mathbf{P}_x)\end{aligned} \tag{4.12}$$

Similarly

$$\begin{aligned}\nabla^2\mathbf{E} &= \mathbf{i}\,\frac{\partial^2 E_x}{\partial z^2}\\ &= \mathbf{i}\left(-\frac{\omega^2 n^2}{c^2}\right)E_x\end{aligned} \tag{4.13}$$

and

$$\frac{\partial^2\mathbf{E}}{\partial t^2} = \mathbf{i}(-\omega^2)E_x \tag{4.14}$$

Writing these results into Eqn (4.7) gives

$$\left(-\frac{\omega^2 n^2}{c^2}\right)E_x = \left(\frac{1}{c^2}\right)(-\omega^2)E_x+\frac{1}{c^2\varepsilon_0}(-\omega^2)P_x+0$$

that is

$$E_x(n^2-1) = P_x/\varepsilon_0 \tag{4.15}$$

For a material where the molecules are well separated (as is the case in a low density gas) we have seen (Chapter 3, Section 3.3) that the polarization is related to the electric field $\mathbf{E}$ by

$$\mathbf{P}_{av} = \varepsilon_0 N\alpha\mathbf{E} \tag{4.16}$$

where $\alpha$ is the molecular polarizability and $N$ is the number of molecules per unit volume. In the present case we are interested in the $x$-components only and we can write

$$P_{x\,av} = \varepsilon_0 N\alpha E_x \tag{4.17}$$

If we are concerned (and this is generally the case) with what happens to the wave as it passes through a large amount of the material (on the molecular scale), we can reasonably assume that any effect of the material will be an average effect of the detailed molecular structure. Then in a calculation involving the polarization of the material we can use the average within the material. Substituting the average value for $P_x$ into our wave equation (Eqn (4.15)), gives the result

$$\boxed{n^2 = 1+N\alpha} \tag{4.18}$$

This is the condition which must be fulfilled if a plane electromagnetic wave is to propagate through a material of the simple type (a low density gas), described by our model. It gives the relationship between the refractive index $n$ and the parameter $\alpha$ in our molecular model. The measurement of the refractive index is a macroscopic measurement involving a large amount of the material and so our use of the average value of the polarization is appropriate.

Since we have obtained this result for a low density gas, we can expect that $n$ will have a value close to unity, that is, the value of $N\alpha$ must be much less than one. Therefore we can write

$$n = (1+N\alpha)^{1/2}$$

and this can be approximated by the first two terms in its binomial expansion, as

$$\boxed{n = 1+\frac{N\alpha}{2}} \tag{4.19}$$

where $\alpha$ is given in terms of our molecular model (see Section 3.1) as

$$\alpha = \frac{q^2/m\varepsilon_0}{(\omega_0^2-\omega^2)+i\gamma\omega}$$

The expression for $\alpha$ shows that the refractive index $n$ will be frequency dependent and can be complex. We give further consideration to these points in Section 4.3.

## 4.2 ELECTROMAGNETIC WAVES IN A DENSE MATERIAL

The expression which we have obtained above for the refractive index applies only to a low density gas. This is because the electric polarization of a dense material is complicated by the contributions to the field which result from separation of charge within the molecules. The problem of calculating the local field within the material has been discussed in Chapter 3 where the Lorentz expression was obtained in the electrostatic case. We can extend the electrostatic result to the case of a wave, provided that the wavelength is sufficiently long to ensure that it covers a large number of molecules. Then, as far as an individual molecule is concerned, the field that it experiences at any time is essentially the same as that experienced by its neighbours.

To obtain the corresponding result for a dense material, the derivation given in Section 4.1 for a low density gas needs to be corrected for the local field at the stage where the relationship between **P** and **E** is introduced. We can continue working from the result shown in Eqn (4.15):

$$E_x(n^2-1) = P_x/\varepsilon_0 \tag{4.20}$$

Now we must use the local field given by the Lorentz equation

$$\mathbf{E}_{local} = \mathbf{E} + \mathbf{P}_{av}/3\varepsilon_0 \tag{4.21}$$

in the expression

$$\mathbf{P}_{av} = N\varepsilon_0\alpha\mathbf{E}_{local} \tag{4.22}$$

since it is the local field which produces the polarization within the material. Then we can write

$$\mathbf{P}_{av} = N\varepsilon_0\alpha[\mathbf{E} + \mathbf{P}_{av}/3\varepsilon_0]$$

that is

$$(1 - N\alpha/3)\mathbf{P}_{av} = N\varepsilon_0\alpha\mathbf{E}$$

$$\mathbf{P}_{av} = \frac{N\varepsilon_0\alpha}{(1 - N\alpha/3)}\mathbf{E} \tag{4.23}$$

The polarization of the material is still linearly related to the field but the constant of proportionality has changed. In the case of our plane wave we can then write

$$P_{x\,av} = \frac{N\varepsilon_0\alpha}{(1 - N\alpha/3)}E_x \tag{4.24}$$

Using this average value for $P_x$ in Eqn (4.20) gives

$$n^2 - 1 = \frac{N\alpha}{(1 - N\alpha/3)}$$

which can be rearranged as

$$\boxed{\frac{3(n^2 - 1)}{n^2 + 2} = N\alpha} \tag{4.25}$$

where the molecular polarizability is again given by

$$\alpha = \frac{q^2/m\varepsilon_0}{(\omega_0^2 - \omega^2) + i\gamma\omega}$$

This result is known as the Clausius-Mosotti equation. Since we have shown that $\varepsilon = n^2$ we can write the Clausius-Mosotti equation in the alternate form in terms of the relative permittivity:

$$3(\varepsilon - 1)/(\varepsilon + 2) = N\alpha$$

We have obtained this expression using the Lorentz equation for the local field in a simple non-polar, homogeneous, isotropic, linear dielectric material. We could expect the Clausius-Mosotti equation to apply to non-polar gases and liquids and in fact the experimental results for such materials are in good agreement with the equation. Notice that for low values of the refractive index $n$, Eqn (4.25) reduces to Eqn (4.18).

## 4.3 THE CONSEQUENCES OF A COMPLEX REFRACTIVE INDEX

We have now obtained two equations

$$n^2 = 1 + N\alpha$$

and

$$\frac{3(n^2-1)}{n^2+2} = N\alpha$$

giving the refractive index $n$ in terms of the molecular polarizability $\alpha$. Which of these equations we use will depend on the density of the material but since each expression contains the term $\alpha$ and since $\alpha$ is given by (see Chapter 3, Section 3.1)

$$\alpha = \frac{q^2/m\varepsilon_0}{(\omega_0^2-\omega^2)+i\gamma\omega}$$

the possibility of a complex refractive index arises through the damping term $i\gamma\omega$. We shall have to return to this topic in greater detail when discussing metals in Chapter 5 and again in Chapter 8, but we should now consider what a complex refractive index means in terms of the propagation of an electromagnetic wave through a material.

The electric field in a plane electromagnetic wave (monochromatic, linearly polarized) is described by

$$\begin{aligned} \mathbf{E} &= \mathbf{i}E_x \\ &= \mathbf{i}E_0 \exp[i\omega(t - zn/c)] \end{aligned}$$

If we write the complex refractive index as

$$n = n_r - in_i \tag{4.26}$$

where $n_r$ and $n_i$ are real, the equation for the electric field becomes

$$\mathbf{E} = \mathbf{i}E_0 \exp(-\omega z n_i/c) \exp[i\omega(t - zn_r/c)]$$

This equation shows that it is the real part of the refractive index which determines the velocity of the wave and so there is no difficulty introduced into our definition of the refractive index (the real part) as the ratio of two velocities. The imaginary term in the complex refractive index causes exponential damping of the amplitude of the wave as it travels through the material. The larger the imaginary component of the refractive index, the more rapidly will the wave be attenuated. This is a direct effect of the damping term $\gamma$ in our molecular model. The molecular polarizability becomes entirely imaginary when $\omega = \omega_0$, that is, when the frequency of the wave corresponds to the resonance frequency of the molecule. At this frequency the energy of the wave is most strongly dissipated by the material. In a real material there will be several such resonant frequencies at which the transmission of radiation will be a minimum.

## 4.4 PHASE VELOCITY AND ENERGY VELOCITY

In Section 4.3 we have seen that the imaginary term in the complex refractive index leads to damping of the wave as it travels but that the velocity of the wave is still determined from the real part of the refractive index. Now we take a closer look at the wave velocity in the relatively simple case of a low density gas.

The refractive index $n$ is approximated by

$$n = 1+\tfrac{1}{2}N\alpha$$

that is

$$n = 1+\left(\frac{Nq^2}{2m\varepsilon_0}\right)\left\{\frac{1}{(\omega_0^2-\omega^2)+i\gamma\omega}\right\}$$

$$= 1+\frac{Nq^2}{2m\varepsilon_0}\left\{\frac{(\omega_0^2-\omega^2)-i\gamma\omega}{(\omega_0^2-\omega^2)^2+\gamma^2\omega^2}\right\}$$

Then, if we write the complex refractive index as

$$n = n_r - in_i$$

the real term in the refractive index is given by

$$n_r = 1+\left(\frac{Nq^2}{2m\varepsilon_0}\right)\left\{\frac{(\omega_0^2-\omega^2)}{(\omega_0^2-\omega^2)^2+\gamma^2\omega^2}\right\} \tag{4.27}$$

and the imaginary term is given by

$$n_i = \frac{\gamma\omega}{(\omega_0^2-\omega^2)^2+\gamma^2\omega^2}\left(\frac{Nq^2}{2m\varepsilon_0}\right) \tag{4.28}$$

Both expressions take on their simplest form at the resonant frequency when $\omega = \omega_0$ and

$$n_r = 1 \tag{4.29}$$

$$n_i = (1/\gamma\omega_0)\left(\frac{Nq^2}{2m\varepsilon_0}\right) \tag{4.30}$$

Whereas $n_i$ is positive at all frequencies, the term $(\omega_0^2-\omega^2)$ becomes negative at frequencies greater than the resonant frequency and the real part of the refractive index becomes less than one. The real part of the refractive index $n_r$ was defined by the ratio of the velocity $c$ of an electromagnetic wave in free space to the velocity $v$ of the wave in the material

$$n_r = c/v$$

If $n_r$ can now become less than one, does this mean that the wave is travelling faster than the velocity of light, in contradiction to Einstein's Special Theory of Relativity? It appears that we must be careful about the velocity of electromagnetic waves in a material. A real difficulty would arise if our result showed that the velocity of the *energy* flow was greater than the velocity $c$. This is not the case in the present situation and to emphasize the fact, the velocity $v$ is called the phase velocity, that is, the speed of the maxima in the sinusoidal wave. More generally, the velocity $v$ is the velocity of 'planes of constant phase' in the wave so that we could, for example, equally well specify the velocity in terms of the speed at which the minima in the electric field pass a fixed point. In Exercise 2.2 it was seen that in free space the energy velocity and the phase velocity both have the value $c$. The complex refractive index for a material means that the phase and energy velocities are no longer equal. In other words, we cannot send a signal (energy) with the phase velocity $v$ (for $v > c$) and so we do not break the physical speed limit $c$. Notice that this complication has only come about through the imaginary part of the refractive index becoming non-zero. The wave is then damped and loses energy as it travels. So if we are observing the minima (or maxima) of the wave in space, the minima will not each have the same magnitude because of the damping. We consider this topic again in Section 4.6 and also in Chapter 5 for guided waves.

## EXERCISE 4.1

The energy velocity is found by dividing the time average value of the Poynting vector by the time average of the energy density. By computing $\mathbf{S}$ and $\mathscr{E}$ from the real parts of the complex expressions for $\mathbf{E}$ and $\mathbf{B}$, find the energy velocity for a simple plane electromagnetic wave travelling in a homogeneous, isotropic, non-conducting medium. Show that as the refractive index approaches that of free space, the energy velocity approaches the value $c$.

### Solution

Taking the wave to be travelling in the positive $z$-direction with the electric field along the $x$-direction, the electric wave is described by

$$\mathbf{E} = \mathbf{i}E_0 \exp[i\omega(t - zn/c)]$$

where the refractive index $n$ is given by

$$n = n_r - in_i$$

From the Maxwell equation

$$\nabla \times \mathbf{E} = -\partial \mathbf{B}/\partial t$$

we can readily show that the associated magnetic field is described by

$$\mathbf{B} = \mathbf{j}(n/c)E$$

The Poynting vector is given by

$$\mathbf{S} = \varepsilon_0 c^2 \mathbf{E} \times \mathbf{B} \ (\mathrm{J\,s^{-1}\,m^{-2}})$$

and the energy density is

$$\mathscr{E} = \frac{\varepsilon_0}{2}(E^2+c^2B^2)\ (\mathrm{J\ m^{-3}})$$

In these expressions we require the real forms of the electric and magnetic fields, which are

$$\mathbf{E} = \mathbf{i}E_0 \exp(-\omega z n_i/c)\cos[\omega(t-zn_r/c)]$$

and

$$\mathbf{B} = \mathbf{j}(E_0/c)\exp(-\omega z n_i/c)\{n_r \cos[\omega(t-zn_r/c)]+n_i \sin[\omega(t-zn_r/c)]\}$$

Then the Poynting vector is

$$\mathbf{S} = \mathbf{k}\varepsilon_0 c^2\left(\frac{1}{c}\right)E_0^2 \exp(-2\omega z n_i/c)\{n_r \cos^2[\omega(t-zn_r/c)] + n_i \sin[\omega(t-zn_r/c)]\cos[\omega(t-zn_r/c)]\}$$

Taking the time average gives

$$\mathbf{S} = \mathbf{k}\varepsilon_0 c E_0^2 \exp(-2\omega z n_i/c)\{n_r(\tfrac{1}{2})\}$$

The energy density is

$$\mathscr{E} = \frac{\varepsilon_0}{2}\left\{E_0^2 \exp(-2\omega z n_i/c)\cos^2[\omega(t-zn_r/c)] + c^2\left[\left(\frac{1}{c}\right)^2 E_0^2 \exp(-2\omega z n_i/c)\{[n_r^2 \cos^2[\omega(t-zn_r/c)] + 2n_r n_i \cos[\omega(t-zn_r/c)]\sin[\omega(t-zn_r/c)] + n_i^2 \sin^2[\omega(t-zn_r/c)]\}\right]\right\}$$

and taking the time average gives

$$\bar{\mathscr{E}} = \frac{\varepsilon_0}{2}\left\{E_0^2 \exp(-2\omega z n_i/c)\{\tfrac{1}{2}\} + E_0^2 \exp(-2\omega z n_i/c)\left[\frac{n_r^2}{2}+\frac{n_i^2}{2}\right]\right\}$$

$$= \frac{\varepsilon_0 E_0^2}{4}\exp(-2\omega z n_i/c)\{1+n_r^2+n_i^2\}$$

The energy velocity is

$$v_{en} = \bar{\mathbf{S}}/\bar{\mathscr{E}} = \mathbf{k}\left\{\frac{2cn_r}{1+n_r^2+n_i^2}\right\}$$

In free space, $n_r = 1$ and $n_i = 0$ which gives $v_{en}$ the value $c$, as expected.

## 4.5 ANOMALOUS DISPERSION

The frequency range close to the resonant frequency is called a 'region of anomalous dispersion' since the 'normal' change in the refractive index with frequency is reversed and d$n$/d$f$ is negative as shown schematically in Fig. 4.1. If a prism of material is used to separate a beam of radiation into the spectrum of its constituent components, then in the region of anomalous dispersion the frequencies will appear in the reverse of their normal order. Note that our model

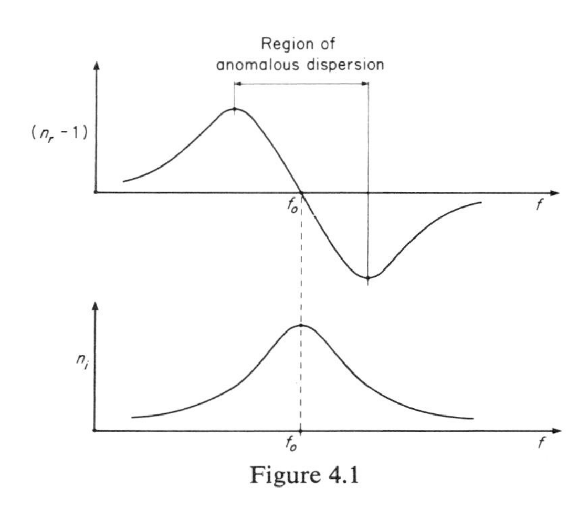


Figure 4.1

shows that in this region the imaginary part of the refractive index will have its maximum value and there will then be very severe energy losses. The variation of $n_i$ is also shown schematically in Fig. 4.1 in the anomalous region. Real materials will have more than one resonant frequency $\omega_0$ because there will be more than one way in which charge can be displaced. Consequently there will be a number of such anomalous regions.

## 4.6 GROUP VELOCITY

We have been considering propagation through a medium which is both dispersive and dissipative. It is *dispersive* because the velocity of propagation of the wave, the phase velocity, depends on the real part of the refractive index and this in turn is frequency dependent. This means that different frequencies propagate at different velocities. The medium is *dissipative* because the non-zero value of the imaginary component of the refractive index leads to damping of the wave as it travels. We have seen that in the present case these two effects are tied together since the imaginary part of the refractive index becomes non-zero as soon as the real part of the refractive index is different from one. We shall later see in Chapter 5 that we can have a medium (a waveguide) which can be dispersive but lossless. It is important, from the point of view of communications, to be able to send signals consisting of more than one frequency and to know how this signal will be affected by dispersion in the medium.

We know that in a dispersive medium two waves of different frequencies will have different propagation velocities. Consider two slightly different waves under these conditions described by $\psi_1$ and $\psi_2$ such that

$$\psi_1 = A\cos[(\omega+\delta\omega)t-(k+\delta k)z] \tag{4.31}$$

and

$$\psi_2 = A\cos[(\omega-\delta\omega)t-(k-\delta k)z] \tag{4.32}$$

where $A$ is a constant. Superimposing the two waves gives (after a lot of algebra!)

$$\psi = \psi_1+\psi_2 = 2A\cos[(\delta\omega)t-(\delta k)z]\cos[\omega t-kz] \tag{4.33}$$

The overall effect is that we have a wave of angular frequency $\omega$ which is modulated by another sinusoidal wave. The result is illustrated in Fig. 4.2.

The 'wave packets' travel with velocity $\delta\omega/\delta k$ while the phase velocity of the basic wave remains $\omega/k$. It is therefore worthwhile to consider an additional velocity, the 'group velocity', $v_g$, which we now define as $d\omega/dk$. Remember that this definition is based on a lossless medium. The meaning of the group velocity is limited to a very narrow frequency band in the case of a dissipative medium because the wave packet will then suffer distortion through differential damping of the different frequency components.

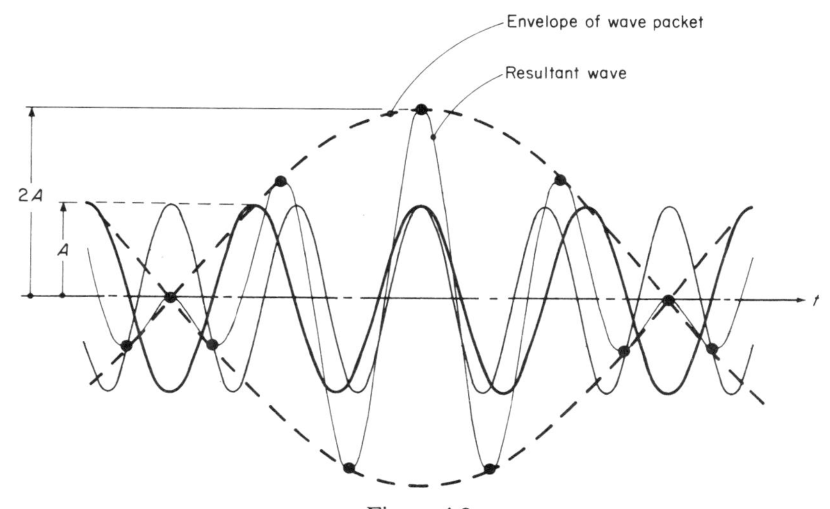


Figure 4.2

## EXERCISE 4.2

Show that the limits of the region of anomalous dispersion are $f = f_0[1 \pm \gamma/2\omega_0]$.

**Solution**

The real part of the refractive index is

$$n_r = 1 + \frac{Nq^2}{2\varepsilon_0 m}\left\{\frac{(\omega_0^2 - \omega^2)}{(\omega_0^2 - \omega^2)^2 + \gamma^2\omega^2}\right\}$$

Differentiating $n_r$ with respect to $\omega$ and equating the derivative to zero will find the turning points on the curve of $n_r$ against frequency. The result is that the turning points occur where

$$2\omega(\omega_0^2 - \omega^2)^2 - 2\gamma^2\omega\omega_0^2 = 0$$

or

$$(\omega_0^2 - \omega^2)^2 = \gamma^2\omega_0^2$$

Therefore

$$\omega_0^2 - \omega^2 = \pm\gamma\omega_0$$
$$\omega^2 = \omega_0^2(1 \pm \gamma/\omega_0)$$

Taking $\gamma/\omega_0 \ll 1$ gives

$$\omega = \omega_0[1 \pm \gamma/2\omega_0]$$

which is the required result

$$f = f_0[1 \pm \gamma/2\omega_0]$$

## 4.7 SUMMARY OF RESULTS

1. The refractive index $n$ of a low density gas is given by

$$n = 1 + N\alpha/2$$

where $N$ is the number of molecules per unit volume and $\alpha$ is the molecular polarizability given by the simple model developed in Chapter 3.

2. The refractive index $n$ of a dense, non-polar, homogeneous, isotropic, dielectric material is given by the Clausius-Mossotti equation

$$\frac{3(n^2 - 1)}{n^2 + 2} = N\alpha$$

This relationship is obeyed by non-polar gases and liquids.

3. The consequence of a complex refractive index

$$n = n_r - in_i$$

is that the imaginary term gives rise to exponential damping of the wave as it travels through a material. The damping is given by a term of the form

$\exp(-\omega z n_i/c)$ which has a maximum value at frequencies close to a resonant frequency for the molecule. In the frequency range close to resonance, the dispersion of electromagnetic radiation within the material is anomalous and the phase velocity of the wave is greater than $c$. In the anomalous region, the material causes the greatest damping of the electromagnetic wave as it propagates through the material.

4. We can define the 'group velocity', $v_g$, as

$$v_g = \frac{\mathrm{d}\omega}{\mathrm{d}k}$$

The meaning of the group velocity is restricted when we consider a medium which is both dispersive and dissipative.

## PROBLEMS

1.
The Lorentz expression for the local field in a dense material was derived for the *static* case in Chapter 3. Now we have extended this result to time varying fields by saying that the wavelength must be sufficiently long to cover an appreciable number of molecules. If we take 10 molecular diameters as the distance defining 'near neighbours' and say that the maximum variation in the field across this group must not exceed 5% of the peak value, obtain an order of magnitude estimate of the upper frequency limit on the basis that small molecules have a diameter of $\sim 2 \times 10^{-10}$ m.

2.
Show that if we consider a dispersive medium well away from the region of anomalous dispersion ($n_i \to 0$) then the group velocity is such that the product of the group and phase velocities is equal to $c^2$.

## FURTHER READING

For further information on wave velocities see Panofsky & Philips and Javid & Brown.

## Chapter 5

# ELECTROMAGNETIC WAVES IN CONDUCTING MATERIALS

The model we take for a metallic medium is based on the theory developed by Sommerfeld, Drude and Lorentz. The model assumes that the electrical properties of a metal are dominated by the conduction (free) electrons (the valency electrons) while the electrons which are tightly bound to the atoms make a comparatively negligible contribution. The conduction electrons are assumed to move freely through the atomic lattice with which they occasionally collide.

This model for a metal can be extended and applied to reflection from a plasma. We conclude this chapter by considering wave propagation through simple waveguides.

## 5.1 A SIMPLE MODEL FOR A METAL

In Chapter 3 we developed a model for the behaviour of the charge within a molecule or atom and we are now going to modify this general model so that we can use it to describe a metal. The model said that the motion of a charge in an atom under the influence of an electric field was described by the force equation

$$qE_x = m\left(\frac{\mathrm{d}^2x}{\mathrm{d}t^2}+\gamma\,\frac{\mathrm{d}x}{\mathrm{d}t}+\omega_0^2x\right) \tag{5.1}$$

where $m$ is the mass of the charge $q$, $\gamma$ is a damping constant for the motion and $\omega_0$ is the natural angular frequency of the charge. It was shown in Chapter 4 that this model gave the refractive index $n$ of a low density material as

$$n^2 = 1+\frac{Nq^2/m\varepsilon_0}{(\omega_0^2-\omega^2)+i\gamma\omega} \tag{5.2}$$

where $N$ is the number of charges ($q$) per unit volume and $\omega$ is $2\pi$ times the measurement frequency.

In the discussion of this result we had seen that it was only suitable for a gas and that for a dense material (a liquid or solid) the result needs to be corrected for local field effects due to the polarization of the molecular dipoles. Such atomic or molecular polarization cannot arise from the free charge in a metal and no correction to the result for $n$ is required to allow for the high density of the metallic medium. Since the free charges are not tightly bound to an atom there is no restoring force proportional to the displacement and these charges cannot have a natural or resonant frequency within the atom and so we modify the general model to represent a metal by writing $\omega_0 = 0$, that is

$$qE_x = m\ddot{x} + m\gamma\dot{x} \tag{5.3}$$

With this modification to the model we can see (from Eqns (5.1) and (5.2)) that the square of the refractive index for a metal is given by

$$n^2 = 1 + \frac{Nq^2/m\varepsilon_0}{-\omega^2 + i\gamma\omega} \tag{5.4}$$

We now attempt to relate the parameters in our microscopic model to the macroscopic quantity, $\sigma$, the electrical conductivity of the metal. For the isotropic conductors that we are considering, the current is proportional to the field and Ohm's law applies and so the current density $\mathbf{J}$ can be written as

$$\mathbf{J} = \sigma\mathbf{E} \tag{5.5}$$

or in one dimension

$$J_x = \sigma E_x \tag{5.6}$$

If the average velocity of the charges moving in the $x$-direction is $\langle v_x \rangle$ then the current is given by

$$J_x = Nq\langle v_x \rangle \tag{5.7}$$

The steady current is controlled by two opposing processes: (i) the field accelerating the charges; (ii) collisions with the lattice slowing the charges. On average these two effects must balance to give a steady current, that is, the average acceleration is zero. For an individual charge we have written (Eqn (5.3))

$$qE_x = m\ddot{x} + m\gamma\dot{x}$$

but in terms of the average velocity we can write

$$qE_x = m\gamma\langle v_x \rangle \tag{5.8}$$

From Eqns (5.6), (5.7), and (5.8)

$$\sigma = \frac{J_x}{E_x} = Nq\frac{\langle v_x \rangle}{E_x} = \frac{Nq^2}{m\gamma}$$

or

$$\sigma = \frac{Nq^2\tau}{m} \quad \text{where} \quad \tau = 1/\gamma \tag{5.9}$$

The constant $\tau$ is introduced for the reciprocal of $\gamma$ to make some of the later equations more convenient.

Equation (5.9) relates the measureable macroscopic quantity $\sigma$ (the electrical conductivity of the metal) to the parameters in our simple model for a metal. We can now examine the model to see what results it gives for the properties of metals.

## 5.2 METALS AT HIGH AND LOW FREQUENCIES

It was shown in Chapter 4 that if a material has a complex refractive index, then, as a result of the imaginary term, the amplitude of an electromagnetic wave decreases exponentially as the wave travels through the medium. If the complex refractive index is written as

$$n = n_r - in_i \tag{5.10}$$

then our expression for the electric field in a plane polarized wave

$$\mathbf{E} = \mathbf{i}E_x = \mathbf{i}E_0 \exp[i\omega(t - nz/c)] \tag{5.11}$$

becomes

$$\mathbf{E} = \mathbf{i}E_0 \exp(-\omega n_i z/c) \exp[i\omega(t - n_r z/c)] \tag{5.12}$$

The imaginary term in the refractive index determines the extent to which the wave is damped as it travels through the medium and we need to evaluate $n_i$ for a metal if we are to study the propagation of electromagnetic waves in metallic media. The distance travelled by the wave before it is appreciably diminished is measured in terms of the skin-depth or penetration depth which is defined as the distance $\delta$ travelled by the wave before its amplitude is reduced by a factor of $e^{-1}$. At this distance the radiated power has been reduced by a factor of $1/e^2$ since the energy in the wave is proportional to the square of the amplitude. From Eqn (5.12) we can write

$$\omega n_i \delta / c = 1$$

then

$$\delta = c/\omega n_i \text{ metres} \tag{5.13}$$

If

$$n = n_r - in_i$$

then

$$n^2 = n_r^2 - n_i^2 - 2in_r n_i \tag{5.14}$$

and our model for a metal has given an expression (Eqn (5.4)) for the square of the refractive index which can be divided into its real and imaginary parts

$$n^2 = 1 + \frac{\sigma/\varepsilon_0}{-\omega^2\tau + i\omega} \quad \text{where} \quad \tau = 1/\gamma$$

$$= 1 + \left(\frac{\sigma}{\varepsilon_0}\right) \frac{(-\omega^2\tau - i\omega)}{(\omega^4\tau^2 + \omega^2)}$$

$$n^2 = 1 - \frac{\sigma\tau/\varepsilon_0}{(\omega^2\tau^2 + 1)} - i\,\frac{\sigma/\varepsilon_0}{(\omega^2\tau^2 + 1)\omega} \tag{5.15}$$

Separately equating the real and imaginary terms in Eqns (5.14) and (5.15) gives

$$n_r^2 - n_i^2 = 1 - \frac{\sigma\tau/\varepsilon_0}{(\omega^2\tau^2+1)} \tag{5.16}$$

and

$$2n_r n_i = \frac{\sigma/\varepsilon_0}{(\omega^2\tau^2+1)\omega}$$

that is

$$n_r = \frac{\sigma/2\varepsilon_0}{(\omega^2\tau^2+1)\omega}\left(\frac{1}{n_i}\right) \tag{5.17}$$

Substituting for $n_r$ in Eqn (5.16) gives

$$\left[\frac{\sigma/2\varepsilon_0}{(\omega^2\tau^2+1)\omega}\left(\frac{1}{n_i}\right)\right]^2 - n_i^2 = 1 - \frac{\sigma\tau/\varepsilon_0}{(\omega^2\tau^2+1)}$$

that is

$$n_i^4 + \left[1 - \frac{\sigma\tau/\varepsilon_0}{(\omega^2\tau^2+1)}\right]n_i^2 - \left[\frac{\sigma/2\varepsilon_0}{(\omega^2\tau^2+1)\omega}\right]^2 = 0 \tag{5.18}$$

To simplify the examination of this equation we write

$$1 - \frac{\sigma\tau/\varepsilon_0}{(\omega^2\tau^2+1)} = k \text{ (say)} \tag{5.19}$$

and

$$\frac{\sigma/2\varepsilon_0}{(\omega^2\tau^2+1)\omega} = \ell \text{ (say)} \tag{5.20}$$

Equation (5.18) then becomes

$$n_i^4 + k n_i^2 - \ell^2 = 0$$

which gives

$$n_i^2 = \frac{-k \pm \sqrt{k^2 + 4\ell^2}}{2}$$

Since this is to be a value for the square of $n_i$, this result must be positive (or zero) and so we obtain $n_i$ as

$$n_i = \pm\left[\frac{-k \pm \sqrt{k^2+4\ell^2}}{2}\right]^{\frac{1}{2}}$$

Only the positive value for $n_i$ corresponds to the case for a damped wave and we take

$$n_i = \left[\frac{-k + \sqrt{k^2+4\ell^2}}{2}\right]^{\frac{1}{2}} \tag{5.21}$$

From Eqn (5.21) we can immediately obtain a qualitative description of the behaviour of metals at high and low frequencies. Consider what happens to $n_i$ as $\omega \to \infty$; we can see that under this condition $\ell \to 0$ while $k \to 1$ and the expression for $n_i$ approaches zero. This means that at very high frequencies the model says that electromagnetic waves will be able to travel through a metal and conversely (see Eqn (5.12)) that at lower frequencies $n_i$ will be finite and there will be appreciable damping of the wave.

To turn this qualitative picture of what happens at high and low frequencies into quantitative information over a wide frequency range means that we must evaluate $n_i$ from Eqn (5.21) for a metal. Taking the data on copper as typical for a metal, the terms in the expression for $n_i$ can be evaluated.

*Data for copper*

| | |
|---|---|
| Electrical conductivity | $= 5.76 \times 10^7\ (\Omega\ \mathrm{m})^{-1}$ |
| Atomic mass | $= 63.5$ |
| Density | $= 8.9 \times 10^3\ \mathrm{kg\ m^{-3}}$ |
| Avogadros' number | $= 6.02 \times 10^{26}$ (kg atomic mass)$^{-1}$ |
| Electronic charge | $= 1.6 \times 10^{-19}$ C |
| Electronic mass | $= 9.11 \times 10^{-31}$ kg |
| Absolute permittivity (free space) | $= 8.85 \times 10^{-12}\ \mathrm{F\ m^{-1}}$ |

Copper is monovalent and since the valency electrons are taken to be the free electrons this means that the number of electrons per unit volume is equal to the number of atoms per unit volume.

$$\text{Number of atoms per m}^3 = \frac{(8.9 \times 10^3) \times (6.02 \times 10^{26})}{63.5}$$

that is

$$N = 8.5 \times 10^{28} \text{ free charges per m}^3.$$

The constant $\tau$ is given by Eqn (5.9)

$$\tau = m\sigma/Nq^2$$

therefore

$$\tau = 2.4 \times 10^{-14} \text{ s (for copper)}$$

Knowing $\sigma$, $\tau$ and $\varepsilon_0$, the terms $k$ and $\ell$ can be written as

$$k = 1 - 1.56 \times 10^5/(\omega^2\tau^2 + 1)$$

and

$$\ell = 3.25 \times 10^8/(\omega^2\tau^2 + 1)\omega$$

For any value of $\omega$, the value of $n_i$ can now be calculated and hence the value of the penetration depth $\delta$. The results are plotted on log scales to accommodate the wide range of values. Figure 5.1 shows the variation of the penetration depth of copper as a function of the angular frequency over part of the frequency range.

At lower frequencies ($\omega < 10^{12}$) the calculation of $n_i$ is completely determined by $\ell$ and the result is that log $\delta$ decreases linearly with log $\omega$. Near the visible range of frequencies the curve flattens and then rises very sharply at ultraviolet and higher frequencies. This is in agreement with the general behaviour of metals which become relatively transparent to ultraviolet, while X-rays can penetrate a considerable thickness of metal. Even in the visible range thin metal films can transmit appreciable amounts of light, though at infrared frequencies their transmission is reduced. At microwave frequencies the penetration depth is still

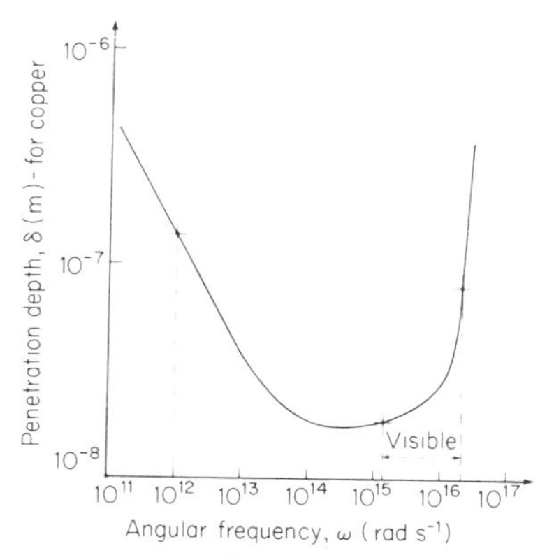


Figure 5.1

sufficiently low so that microwave waveguides can be fabricated from relatively poor conductors coated with a thin, highly-conducting layer of silver which is sufficient to contain the wave in what is, in effect, a highly conducting guide. Our model gives quite a good description of the general behaviour of metals but it cannot describe the detailed behaviour of particular metals, especially at very high frequencies where there are effects well outside the limitations of this simple model.

The simplicity of the calculation of $n_i$ at values of $\omega < 10^{12}$ suggests that we could simplify the expression for the penetration depth $\delta$ at relatively low frequencies. If we say that the term $(\omega^2\tau^2 + 1)$ is to be approximately one

$$\omega^2\tau^2 + 1 \simeq 1 \tag{5.22}$$

then that is equivalent to writing

$$\omega \ll 1/\tau \tag{5.23}$$

and from Eqn (5.22) we can immediately write the expression for $n_i$ as

$$n_i = \left[\frac{\left(\dfrac{\sigma\tau}{\varepsilon_0} - 1\right) + \sqrt{\left(\dfrac{\sigma\tau}{\varepsilon_0} - 1\right)^2 + \left(\dfrac{\sigma}{\varepsilon_0\omega}\right)^2}}{2}\right]^{\frac{1}{2}} \tag{5.24}$$

From the condition given in Eqn (5.23) it is clear that

$$\left(\frac{\sigma\tau}{\varepsilon_0} - 1\right) \ll \left(\frac{\sigma}{\varepsilon_0\omega}\right)$$

and the expression for $n_i$ becomes

$$n_i = \sqrt{\frac{\sigma}{2\varepsilon_0\omega}} \tag{5.25}$$

from which the penetration depth is now

$$\delta = \sqrt{\frac{2\varepsilon_0 c^2}{\sigma\omega}} \text{ metres} \tag{5.26}$$

where $\omega \ll 1/\tau$.

The value of $\tau$ for copper (calculated above) is $2.4 \times 10^{-14}$ s and this gives the upper frequency limit for the use of Eqn (5.26) as much less than $[1/(2.4 \times 2\pi)] \times 10^{14}$ Hz.

We conclude that for a good conductor, Eqn (5.26) can be used to calculate the penetration depth for frequencies of less than $\sim 10^{11}$ Hz.

## 5.3 REFLECTION FROM A PLASMA

We can apply our model for a metal to the case of a highly ionized gas (a plasma). The mass of an atomic ion is so much greater than the mass of an electron that the electrical properties of the plasma are dominated by the movement of the free electrons. The ionized layer (the ionosphere) surrounding the Earth is important in radio communications since it can reflect such waves back to Earth. This allows reception of signals from distant stations which are too far apart to be on the line of sight. The signal hops around the Earth by multiple reflections between the Earth and the ionosphere. The conditions for reflection from such an ionized layer can be found from the real part of its refractive index and Snell's law. The real part of the refractive index is obtained in an exactly similar way to that used in Section 5.2 for the imaginary term in the refractive index of a metal.

## EXERCISE 5.1

Show that at high frequencies ($\omega \gg 1/\tau$, that is, $\omega \gg \gamma$) the real part $n_r$ of the refractive index of a metal or highly conducting plasma is given by

$$n_r = \sqrt{1 - \frac{\sigma\tau}{\varepsilon_0(\omega^2\tau^2 + 1)}} = \sqrt{1 - \frac{Nq^2}{m\varepsilon_0\omega^2}}$$

**Solution**

From Eqns (5.16) and (5.17) we obtain the equation determining $n_r$ as

$$n_r^4 - \left[1 - \frac{\sigma\tau}{\varepsilon_0(\omega^2\tau^2 + 1)}\right] n_r^2 - \left[\frac{\sigma}{2\varepsilon_0\omega(\omega^2\tau^2 + 1)}\right]^2 = 0$$

which we can write as

$$n_r^4 - k n_r^2 - \ell^2 = 0$$

then

$$n_r^2 = \frac{k \pm \sqrt{k^2 + 4\ell^2}}{2}$$

but $n_r^2$ must be positive and so we take

$$n_r^2 = \frac{k + \sqrt{k^2 + 4\ell^2}}{2}$$

and

$$n_r = \left[\frac{k + \sqrt{k^2 + 4\ell^2}}{2}\right]^{\frac{1}{2}}$$

Comparing $k$ and $\ell$ we can see that if $\omega \ll 1/\tau$ then the term $\ell$ dominates the calculation of $n_r$ and we obtain

$$\begin{aligned} n_r &= \sqrt{\ell} \\ &= \sqrt{\frac{\sigma}{2\varepsilon_0 \omega(\omega^2 \tau^2 + 1)}} \\ &= \sqrt{\frac{Nq^2}{2m\gamma\varepsilon_0 \omega(1 + \omega^2/\gamma^2)}} \\ &= \sqrt{\frac{Nq^2 \gamma}{2m\varepsilon_0 \omega(\gamma^2 + \omega^2)}} \end{aligned}$$

Then, under the condition $\omega \ll \gamma$, that is, $\omega \ll 1/\tau$, we can write

$$n_r = \sqrt{\frac{Nq^2}{2m\varepsilon_0 \omega \gamma}} \tag{5.26}$$

This then is the real part of the refractive index at *low* frequencies. At high frequencies where $\omega \gg 1/\tau$, that is, $\omega \gg \gamma$, the term $k$ dominates the calculation of $n_r$ and we obtain

$$\begin{aligned} n_r &= \sqrt{k} \\ &= \sqrt{1 - \frac{\sigma\tau}{\varepsilon_0(\omega^2 \tau^2 + 1)}} \\ &= \sqrt{1 - \frac{Nq^2}{m\varepsilon_0 \gamma^2 (1 + \omega^2/\gamma^2)}} \\ &= \sqrt{1 - \frac{Nq^2}{m\varepsilon_0(\gamma^2 + \omega^2)}} \end{aligned}$$

Under the high frequency condition $\omega \gg \gamma$, we can write

$$n_r = \sqrt{1 - \frac{Nq^2}{m\varepsilon_0\omega^2}} \tag{5.27}$$

The expression for $n_r$ at high frequencies shows that the real part of the refractive index will be less than one and (from our discussion in Chapter 4, Section 4.4) the exponentially damped wave will have a phase velocity greater than $c$. The

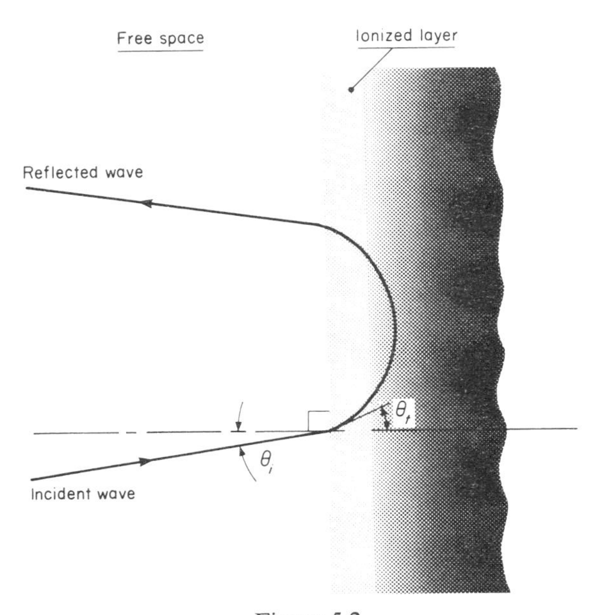


Figure 5.2

conditions for reflection of the wave are complicated by the lack of a sharp boundary to the ionosphere. The degree of ionization varies in the layer, which means that, as $N$ changes, the value of $n_r$ will depend on position. Under these conditions, reflection will correspond to a gradual turning of the incident wave until it comes back out of the ionized layer. At each point, Snells' law (derived in Section 6.3) relates the local value of $n_r$ to the angle of incidence $\theta_i$ and the angle of transmission $\theta_t$ as shown in Fig. 5.2:

$$n_r \sin \theta_t = \sin \theta_i$$

Then if $n_r$ decreases because $N$ is increasing, $\sin \theta_t$ and therefore $\theta_t$ must increase. We can obtain the condition that the wave should be returned by saying that if $n_r$ reaches a value equal to $\sin \theta_i$, then $\sin \theta_t$ must be one and $\theta_t = 90°$. Then the wave is just reflected if

$$\sqrt{1-\frac{Nq^2}{m\varepsilon_0\omega^2}} = \sin\theta_i$$

$$1-\frac{Nq^2}{m\varepsilon_0\omega^2} = \sin^2\theta_i$$

$$\frac{Nq^2}{m\varepsilon_0\omega^2} = \cos^2\theta_i$$

or

$$\boxed{\omega = \frac{1}{\cos\theta_i}\sqrt{\frac{Nq^2}{m\varepsilon_0}}} \qquad (5.28)$$

For normal incidence $\theta_i = 0$ and $\cos\theta_i = 1$ which gives

$$\omega = \sqrt{\frac{Nq^2}{m\varepsilon_0}}$$

This gives the maximum frequency (the critical frequency $f_c$) for which reflection will occur at normal incidence, as

$$\boxed{f_c = \frac{1}{2\pi}\sqrt{Nq^2/m\varepsilon_0}} \qquad (5.29)$$

Since $N$ depends on the degree of ionization of the layer and this will have a value which depends on the amount of radiation from the Sun, the value of $f_c$ can vary both from place to place around the Earth and from time to time. It is an everyday experience that radio reception from distant stations can change dramatically between night and day. It should be noted that the real situation in the ionosphere is very complex. The type of ion present and the ion number density both vary with height above the Earth's surface in such a way that we speak of different distinct ionized layers. These layers fluctuate in position and in their effect depending on the variation in the radiation from the Sun. The cyclic variation in sunspot activity leads to marked changes in global radio propagation patterns.

## 5.4 GUIDED WAVES

We have examined what happens to a plane wave as it propagates through a metallic medium and the results show that the imaginary term in the refractive index of the metal gives exponential damping of the wave through the term $\exp[-\omega n_i z/c]$. The complete expression for $n_i$ is complicated but we have found that if $\sigma/2\varepsilon_0\omega(\omega^2\tau^2+1)$ is zero then $n_i$ is zero and the wave penetrates the metal, but for an *ideal* metal $\sigma \rightarrow \infty$ and the wave cannot propagate through the metal. We now consider the possibilities for getting our plane wave through a 'hole' in such an ideal metal. The problem is simplified by considering the 'hole' or 'waveguide' to be along the $z$-axis and by saying that it is to have a rectangular cross section as shown in Fig. 5.3.

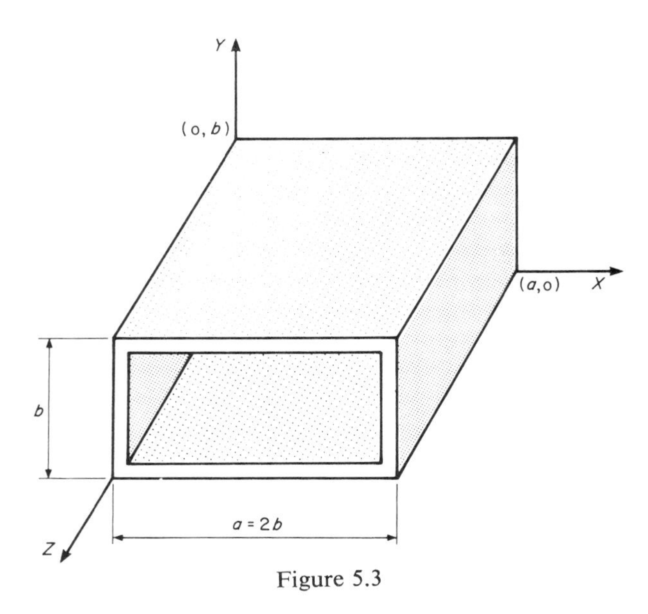


Figure 5.3

We must now decide how we are going to describe the wave. A plane monochromatic wave can be described by the expression

$$\mathbf{E} = \mathbf{E}_0 \exp[i(\omega t - kz)]$$

where the wave is travelling along the $z$-direction with a phase velocity $\omega/k$ and the direction of the electric field is given by the constant vector $\mathbf{E}_0$. In this case we can specify $\mathbf{E}_0$ to be in the $y$-direction.

As usual, the description of the medium is important and in this case the medium (the 'hole' or 'waveguide') can be taken to be free space with $\rho = 0$ and $\mathbf{J} = 0$, that is, with zero charge density and zero current density. The boundary of the waveguide must also be considered; here we have an ideal metal with an electrical conductivity approaching infinity and, from the discussion given above, the properties of the metallic boundary are not frequency dependent because they are dominated by the condition $\sigma \rightarrow \infty$. The boundary conditions are then the same as those for a static field. This means that the electric field must be normal to the boundary and there can be no tangential component since, if the conductivity approaches infinity, the charges will immediately move to reduce any electric field to zero within the perfect conductor. So how does this leave our plane wave? We must check what happens as we move in the $x$-$y$ plane because such variations bring us into contact with the boundary. As we move in the $y$-direction (from the centre of the guide, say) everything is all right because the electric field is pointing in the $\pm y$-direction and so it strikes the boundary at right angles. As we move in the $x$-direction we find that, if the wave is described

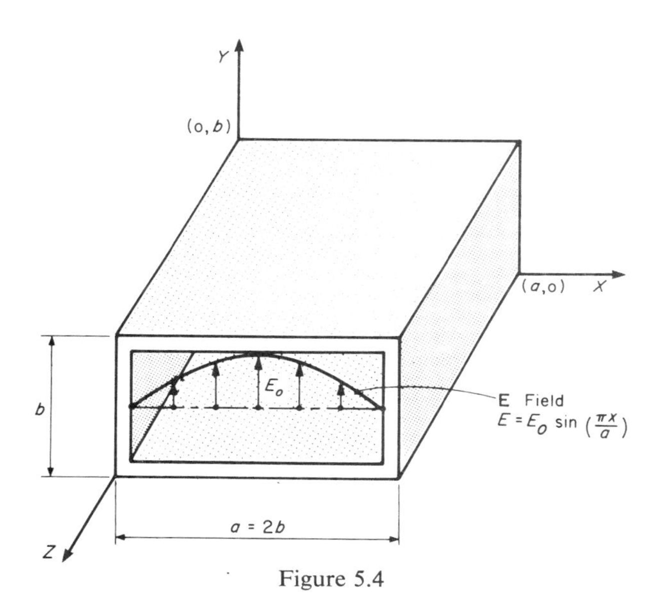


Figure 5.4

by Eqn (5.30), the electric field is parallel to the boundary and this is *not* acceptable from the boundary conditions. Somehow we must arrange the wave so that the electric field is zero at the boundary limits of $x$. One way in which this could be achieved and which would still allow us to have a wave in the rest of the guide is if the field intensity were to vary with $x$ as $\sin(\pi x/a)$. This ensures that there is zero field at $x = 0$ and $x = a$ as shown in Fig. 5.4. Of course the same result would be obtained with the more general function $\sin(n\pi x/a)$ where $n = 0, 1, 2 \ldots$ but for the moment we take the simplest case of $n = 1$ and write the equation for the plane wave as

$$\mathbf{E} = \mathbf{j}E_0 \sin(\pi x/a) \exp[i(\omega t - kz)] \tag{5.31}$$

where $\mathbf{j}$ is the unit vector in the $y$-direction.

Now that the wave satisfies the boundary conditions we can check to see if it satisfies the Maxwell equations in the free space region of the guide. In Chapter 2 it was shown that the Maxwell equations in free space gave the wave equation for the electric field as

$$\nabla^2\mathbf{E} - \frac{1}{c^2}\frac{\partial^2\mathbf{E}}{\partial t^2} = 0 \tag{5.32}$$

that is

$$\frac{\partial^2 \mathbf{E}}{\partial x^2}+\frac{\partial^2 \mathbf{E}}{\partial y^2}+\frac{\partial^2 \mathbf{E}}{\partial z^2}-\frac{1}{c^2}\frac{\partial^2 \mathbf{E}}{\partial t^2} = 0 \tag{5.33}$$

In the present case, **E** has only a *y*-component and substitution for **E** in this wave equation gives

$$E_0\left(\frac{\pi}{a}\right)^2\left[-\sin\left(\frac{\pi x}{\alpha}\right)\right] \exp[i(\omega t-kz)]+0$$

$$+E_0\left[\sin\left(\frac{\pi x}{a}\right)\right](-k^2)\exp[i(\omega t-kz)]$$

$$-\frac{1}{c^2}E_0\left[\sin\left(\frac{\pi x}{a}\right)\right](-\omega^2)\exp[i(\omega t-kz)] = 0$$

that is

$$-\left(\frac{\pi}{a}\right)^2-k^2+\frac{\omega^2}{c^2} = 0$$

$$\boxed{k = \pm\sqrt{\left(\frac{\omega}{c}\right)^2-\left(\frac{\pi}{a}\right)^2}} \tag{5.34}$$

If $k$ satisfies this condition then the wave described by Eqn (5.31) satisfies the wave equation (Eqn (5.32)) and the wave can propagate through the waveguide. The $\pm$ in the solution simply allows the wave to travel in either direction. The magnitude of $k$ will determine the phase velocity, $v$, of the wave in the guide by

$$v = \omega/k = \omega\Big/\sqrt{\left(\frac{\omega}{c}\right)^2-\left(\frac{\pi}{a}\right)^2} \tag{5.35}$$

Since $k$ is given by the square root of the difference in two squares, it can be real or imaginary. The imaginary value occurs if $(\omega/c) < (\pi/a)$, and $k$ becomes

$$k = \pm i\sqrt{\left(\frac{\pi}{a}\right)^2-\left(\frac{\omega}{c}\right)^2}$$

$$= \pm ik' \text{ (say)}$$

Under this condition Eqn (5.31) describing the wave in the guide, becomes

$$E = E_0\left[\sin\left(\frac{\pi x}{a}\right)\right] \exp(-k'z)\exp(i\omega t)$$

which represents an exponentially damped electric field which varies sinusoidally with time. This result is *not* the same as that for propagation through a non-ideal metal because there is no real part for $k$ and this means that the wave does not travel forward with time as a sinusoidal wave in space. The difference is illustrated in Fig. 5.5 which shows the instantaneous spatial variation of the magnitude of the electric field in an ideal waveguide for $f < c/2a$ and in a non-ideal metal.

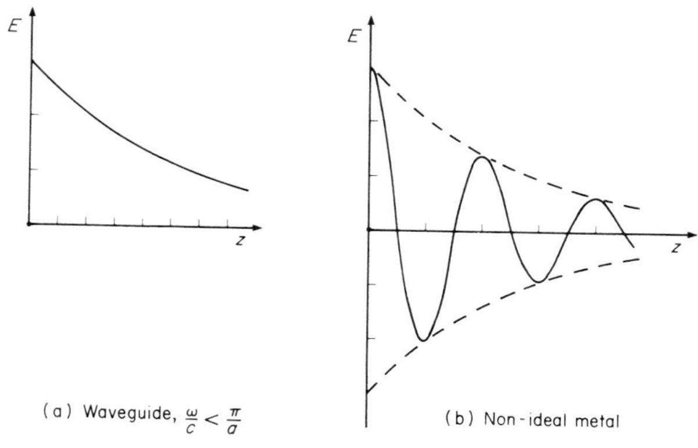


Figure 5.5

We conclude that if $\omega < \pi c/a$ there cannot be propagation through the wave guide and that this frequency corresponds to a cut off frequency, $f_c$, for the guide where $f_c = c/2a$. Now consider the magnetic field **B** in the guide.

The second Maxwell equation states that

$$\nabla \times \mathbf{E} = -\partial \mathbf{B}/\partial t$$

and in the present case

$$\mathbf{E} = \mathbf{j}E_0\left[\sin\left(\frac{\pi x}{a}\right)\right] \exp[i(\omega t - kz)]$$

$$= \mathbf{j}E \tag{5.36}$$

Therefore

$$\nabla \times \mathbf{E} = \begin{vmatrix} \mathbf{i} & \mathbf{j} & \mathbf{k} \\ \dfrac{\partial}{\partial x} & \dfrac{\partial}{\partial y} & \dfrac{\partial}{\partial z} \\ 0 & E & 0 \end{vmatrix}$$

$$= \mathbf{i}\left(-\frac{\partial E}{\partial z}\right) - \mathbf{j}(0) + \mathbf{k}\left(\frac{\partial E}{\partial x}\right)$$

$$= \mathbf{i}(ikE) + \mathbf{j}(0) + \mathbf{k}\left\{\frac{\pi}{a}\left[\cos\left(\frac{\pi x}{a}\right)\right] E/\sin\left(\frac{\pi x}{a}\right)\right\}$$

The magnetic field is then found by partially integrating with respect to time.

Since we are concerned with a wave solution only, the integration constant (a function of the space coordinates only) is of no interest and can be set to zero.

$$\therefore \qquad \mathbf{B} = \mathbf{i}\left(-\frac{ik}{i\omega}\right)E+\mathbf{j}(0)+\mathbf{k}\left\{\left(-\frac{\pi}{a}\right)\left[\cot\left(\frac{\pi x}{a}\right)\right]E/i\omega\right\}$$

or

$$\mathbf{B} = \mathbf{i}\left(-\frac{k}{\omega}\right)\left[\sin\left(\frac{\pi x}{a}\right)\right]E_0\exp[i(\omega t-kz)]$$
$$+\mathbf{j}(0)$$
$$+\mathbf{k}\left(\frac{i\pi}{a\omega}\right)\left[\cos\left(\frac{\pi x}{a}\right)\right]E_0\exp[i(\omega t-kz)] \qquad (5.37)$$

where we are retaining only terms which vary with $t$ since these are the only terms of interest in the wave. The **k** component of **B** is multiplied by $i$ and we need to remember that the real field is described by the *real* part of the complex vector field. This means that the **k** component of **B** is shifted in phase. The fact that the magnetic field has a **k** component is interesting since this is the first case where we have found a wave which has a field with a component along the direction of propagation. From the real parts of Eqns (5.36) and (5.37) the arrangement of the electric and magnetic fields in the guide can be sketched. This is left as an exercise and the result is shown in Appendix 5.1 at the end of this chapter.

Equation (5.34) shows that the propagation constant $k$ can be either real or imaginary and if $k$ is imaginary (for $\omega < \pi c/a$) then the wave cannot propagate in the guide. If $k$ is real then Eqn (5.35) shows that the phase velocity of the wave is greater than $c$. Such a situation was considered in Chapter 4, Section 4.4, for a wave in a dielectric medium but the present case is not quite the same because there is no attenuation of the wave as it travels through our ideal wave guide. In the discussion given in Section 4.4 we saw that the phase velocity had a rather limited meaning and we now calculate the more important quantity, the velocity with which energy can travel through the guide when the wave is described by Eqns (5.36) and (5.37).

The Poynting vector **S** gives the rate at which energy is flowing per unit area at a point in an electromagnetic field where

$$\mathbf{S} = \varepsilon_0 c^2 \mathbf{E}\times\mathbf{B}$$

and where the direction of **S** gives the direction of flow of energy. The real electric and magnetic fields are obtained from Eqns (5.36) and (5.37) as

$$\mathbf{E} = \mathbf{j}E_0\left[\sin\left(\frac{\pi x}{a}\right)\right][\cos(\omega t-kz)] \qquad (5.38)$$

$$\mathbf{B} = \mathbf{i}\left(-\frac{k}{\omega}\right)E_0\left[\sin\left(\frac{\pi x}{a}\right)\right][\cos(\omega t-kz)]-\mathbf{k}\left(\frac{\pi}{a\omega}\right)E_0\left[\cos\left(\frac{\pi x}{a}\right)\right][\sin(\omega t-kz)] \qquad (5.39)$$

The Poynting vector is

$$\mathbf{S} = \varepsilon_0 c^2 \begin{vmatrix} \mathbf{i} & \mathbf{j} & \mathbf{k} \\ 0 & \left\{ E_0 \left[\sin\left(\frac{\pi x}{a}\right)\right] \times \cos(\omega t - kz) \right\} & 0 \\ \left\{ \left(-\frac{k}{\omega}\right) E_0 \left[\sin\left(\frac{\pi x}{a}\right)\right] \times \cos(\omega t - kz) \right\} & 0 & \left\{ \left(-\frac{\pi}{a\omega}\right) E_0 \left[\cos\left(\frac{\pi x}{a}\right)\right] \times \sin(\omega t - kz) \right\} \end{vmatrix}$$

$$\mathbf{S} = \varepsilon_0 c^2 E_0^2 \left\{ \mathbf{i}\left(-\frac{\pi}{a\omega}\right)\left(\frac{1}{4}\right)\left[\sin\left(\frac{2\pi x}{a}\right)\right] \sin[2(\omega t - kz)] + \mathbf{k}\left(\frac{k}{\omega}\right) \sin^2\left(\frac{\pi x}{a}\right) \cos^2(\omega t - kz) \right\}$$

It is interesting to see that the Poynting vector shows that the energy does not simply travel along the guide axis. This is because there is an **i** term in addition to the **k** component in **S**, but we now want to calculate the flow *through* the guide, that is, the flow in the $z$-direction. The rate at which energy flows across an element $dx\,dy$ of the cross section of the guide is given by $\mathbf{S}\cdot\mathbf{n}\,dx\,dy$ where $\mathbf{n}$ is the unit normal to the element of area. Since we wish to calculate the flow through the guide we take $\mathbf{n} = \mathbf{k}$ and

$$\mathbf{S}\cdot\mathbf{n}\,dx\,dy = \varepsilon_0 c^2 E_0^2 \left(\frac{k}{\omega}\right)\left[\sin^2\left(\frac{\pi x}{a}\right)\right][\cos^2(\omega t - kz)]\,dx\,dy$$

The total rate of flow of energy $du/dt$ through the cross section of the guide is found by integrating over the cross section, between the limits for $x$ and $y$

$$\frac{du}{dt} = \int_0^b \int_0^a \varepsilon_0 c^2 E_0^2 \left(\frac{k}{\omega}\right)\left[\sin^2\left(\frac{\pi x}{a}\right)\right][\cos^2(\omega t - kz)]\,dx\,dy$$

which (see Appendix B) gives the flow through a cross section as

$$\frac{du}{dt} = \varepsilon_0 c^2 \left(\frac{k}{\omega}\right) E_0^2 \left(\frac{ab}{2}\right) \cos^2(\omega t - kz) \text{ J s}^{-1}$$

Taking the time average gives

$$\overline{\frac{du}{dt}} = \varepsilon_0 c^2 \left(\frac{k}{\omega}\right) E_0^2 \left(\frac{ab}{2}\right)\left(\frac{1}{2}\right) \text{ J s}^{-1} \tag{5.40}$$

The energy density $\mathscr{E}$ at a point in the waveguide is given by (see Chapter 1, Section 1.7)

$$\mathscr{E} = \frac{\varepsilon_0}{2}(\mathbf{E}^2 + c^2\mathbf{B}^2) \text{ J m}^{-3}$$

which from Eqns (5.38) and (5.39) is

$$\mathscr{E} = \frac{\varepsilon_0}{2}\left\{E_0^2\left[\sin^2\left(\frac{\pi x}{a}\right)\right]\cos^2(\omega t-kz)\right.$$

$$+c^2\left[\left(\frac{k}{\omega}\right)^2 E_0^2\left[\sin^2\left(\frac{\pi x}{a}\right)\right]\cos^2(\omega t-kz)\right.$$

$$\left.\left.+\left(\frac{\pi}{a\omega}\right)^2 E_0^2\left[\cos^2\left(\frac{\pi x}{a}\right)\right]\sin^2(\omega t-kz)\right]\right\}$$

Integrating $\mathscr{E}$ over the cross-section of the guide and taking the time average (the values of the appropriate integrals are given in Appendix B) gives the average energy per unit length of the guide as

$$\bar{\mathscr{E}}_{av} = \frac{\varepsilon_0}{2}E_0^2\left\{\frac{ab}{2}\left(\frac{1}{2}\right)+c^2\left[\left(\frac{k}{\omega}\right)^2\left(\frac{ab}{2}\right)\left(\frac{1}{2}\right)+\left(\frac{\pi}{a\omega}\right)^2\left(\frac{ab}{2}\right)\left(\frac{1}{2}\right)\right]\right\}$$

$$= \frac{\varepsilon_0}{8}E_0^2 ab\left\{1+\frac{c^2k^2}{\omega^2}+\frac{c^2\pi^2}{a^2\omega^2}\right\}$$

but from Eqn (5.34)

$$k^2 = \left(\frac{\omega}{c}\right)^2-\left(\frac{\pi}{a}\right)^2$$

and therefore

$$\bar{\mathscr{E}}_{av} = \frac{\varepsilon_0}{4}E_0^2 ab \text{ J m}^{-1} \tag{5.41}$$

If we divide the average rate at which energy is flowing through the guide $\overline{du}/dt$ by the average energy per unit length of the guide $\bar{\mathscr{E}}_{av}$ we obtain the average rate of flow of energy through the guide: the energy velocity, $v_{en}$ which is

$$v_{en} = \frac{c^2k}{\omega} \text{ m s}^{-1} \tag{5.42}$$

This can also be written as

$$v_{en}v = c^2 \tag{5.43}$$

Although the phase velocity $v$ is greater than $c$, the energy velocity $v_{en}$ can never be greater than $c$. Notice that these velocities depend on frequency and so our ideal waveguide is dispersive since different frequencies will travel with different velocities. The wave we have considered is only one of the many types of wave which can be propagated through a wave guide. The electric field was transverse to the direction of propagation (though the result for **B** shows that it has a component in the $z$-direction) and this type of wave is called a transverse electric wave or TE wave. If we had started by specifying that the **B** field was in the

$y$-direction then we would have had a transverse magnetic wave, that is a TM wave.

The problem was further simplified by taking $n = 1$ in the general function $\sin(n\pi x/a)$ which was introduced to satisfy the boundary condition. If we had retained $n$ we would have found

$$k = \pm\sqrt{\left(\frac{\omega}{c}\right)^2 - \left(\frac{n\pi}{a}\right)^2} \tag{5.44}$$

in which case the cut off frequency is

$$f_c = \frac{nc}{2a} \tag{5.45}$$

where higher values of $n$ give higher cut off frequencies.

An even more general type of wave could have been propagated by taking the electric field to have $x$ and $y$ components which would have introduced $x$ and $y$ boundary conditions. This gives rise to an integer $m$ in addition to $n$ and the general wave could be specified as $TE_{n,m}$. The particular case we have taken would be described as $TE_{1,0}$. In general the cut off frequency for a particular mode is given by

$$f_{n,m} = \sqrt{\left(\frac{mc}{2b}\right)^2 + \left(\frac{nc}{2a}\right)^2} \tag{5.46}$$

We have only considered a wave guide constructed from an ideal metal where there are no energy losses from the guide. In Section 5.2, the relationship for the penetration depth $\delta$ shows that, in the frequency range for which waveguides are useful, the penetration depth is small and a thin layer of a good conductor deposited on the inside surface of the guide minimizes the losses.

## 5.5 SUMMARY OF RESULTS

1. At very high frequencies electromagnetic waves can penetrate metals but at lower frequencies the waves are severely damped.

2. For low frequencies, that is, $f \ll Nq^2/m\sigma$, (where $\sigma$ is the conductivity) the penetration depth $\delta$ is given by

$$\delta = \sqrt{\frac{2\varepsilon_0 c^2}{\sigma\omega}}$$

3. A plasma behaves very much like a metal and will reflect high frequencies up to the critical frequency $f_c$ (for normal incidence) given by

$$f_c = \frac{1}{2\pi}\sqrt{\frac{Nq^2}{m\varepsilon_0}}$$

4. Electromagnetic waves can travel without loss through an ideal waveguide. For a rectangular waveguide (dimensions $a \times b$) there is a lower limit to the frequency which can be propagated given by

$$f_c = \frac{c}{2a}$$

5. The velocity of energy flow $v_{en}$ through the guide is related to the phase velocity $v$ of the wave by

$$v_{en} v = c^2$$

Although $v$ can be greater than $c$, $v_{en}$ can never exceed $c$.

6. The ideal waveguide, though lossless, is dispersive.

## APPENDIX 5.1

Sketch of **E** and **B** fields in a rectangular waveguide.

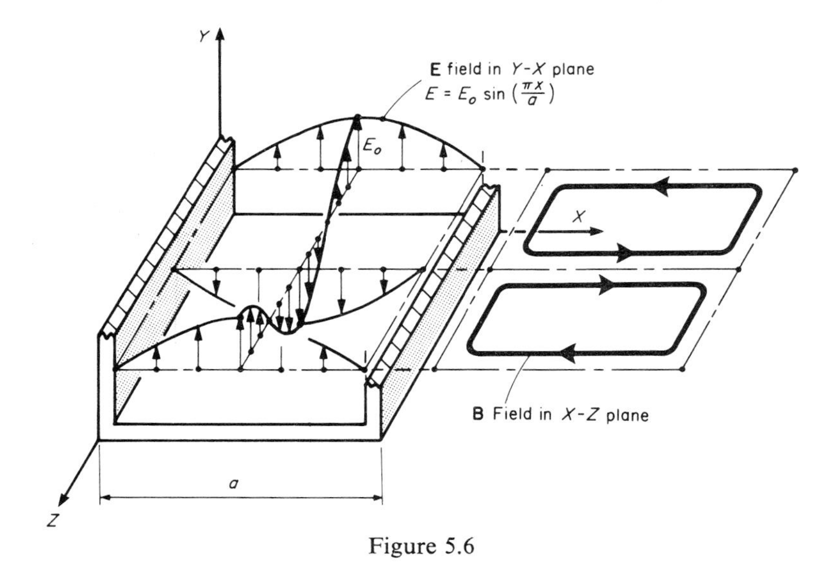


Figure 5.6

## PROBLEMS

1.
Estimate the penetration depth for silver at the microwave frequency of 10 GHz, given that its electrical conductivity is $6.17 \times 10^7$ $(\Omega\text{m})^{-1}$ and taking silver to be monovalent.

2.
The electrical conductivity of seawater is approximately $4.6\ \Omega^{-1}\ \mathrm{m}^{-1}$. Calculate the frequency for which the penetration depth of seawater is 2 m. At this frequency, what distance can be travelled through seawater before the power is reduced to 0.0001 % of its initial value?

3.
Calculate the critical frequency for reflection at normal incidence for (a) the ionosphere, assuming an electron number density ($N$) of $\sim 10^{12}\ \mathrm{m}^{-3}$, (b) for a metal—data for copper is given in Section 5.2.

4.
Calculate the minimum cut off frequency for the $TE_{10}$ mode for an ideal rectangular waveguide whose dimensions are $2.5 \times 5.0$ cm. Note: The two to one ratio in the dimensions of a rectangular waveguide is employed to prevent the propagation of unwanted modes.

5.
Calculate the maximum wavelength that will propagate in the $TE_{10}$ mode for an ideal rectangular waveguide whose internal dimensions are $1\ \mathrm{cm} \times 2\ \mathrm{cm}$. Calculate the energy velocity and the phase velocity through the guide at twice the calculated cut off frequency. Are the results physically acceptable? Does the wave lose energy as it travels through this waveguide?

## FURTHER READING

A discussion of the optical properties of metals is given by Andrews and by Jenkins & White. Further information on waveguides is provided by Paris & Hurd, and by Magid. See Papas, Clemmow, and Javid & Brown for a discussion of plasmas. Stratton discusses wave velocities; see also Panofsky & Philips. For a concise description of the ionosphere see the Encyclopedia of Science & Technology, 254–256, vol. 7, McGraw-Hill (1966).

# Chapter 6

# REFLECTION AND REFRACTION

So far we have been discussing the Maxwell equations in different media (free space, dielectric and metallic media) but we have not considered how an electromagnetic wave could get started (this we leave until Chapter 8), or what happens to the wave as it passes from one medium to another. It is the latter problem that we now consider for the case of a plane electromagnetic wave striking the boundary between two ideal non-conducting media. We need to equip ourselves with a set of conditions (the boundary conditions) which apply to the electric and magnetic fields on either side of the boundary between two homogeneous, isotropic, dielectric media. We also need to extend our mathematical description of an electromagnetic wave so that we can allow it to change direction without slipping out of our mathematical grasp.

## 6.1 THE BOUNDARY CONDITIONS

The problem is illustrated in Fig. 6.1 which shows the boundary and the directions in which the waves are travelling.

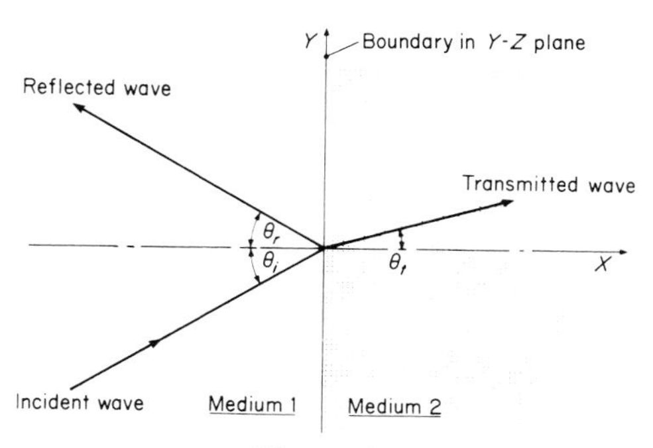


Figure 6.1

For simplicity we consider a plane boundary between medium 1 and medium 2, with the incident wave travelling through medium 1 to the boundary, and we arrange the coordinate system so that the boundary is in the $y$-$z$ plane with the

$x$-axis perpendicular to the boundary. This means that to experience the boundary we must travel in the $x$-direction. If we do not move so that we have an $x$-component in our velocity we never find the boundary. For example, if we allow only the time to vary, then we never experience any changes due to the boundary, because a time variation alone does not take us across the boundary. The boundary can be regarded as a small region of continuous change from medium 1 to medium 2. The idea of a mixed region is not unrealistic when an interface such as, for example, the glass/air boundary is considered; the surface of the glass will contain an adsorbed layer of air so that there is a region where both materials are present. Maxwell's equations are to apply to medium 1, of refractive index $n_1$, say, and to medium 2, of refractive index $n_2$, say, as well as to the boundary region.

In Chapter 3 we wrote the Maxwell equations for an isotropic dielectric medium as

$$\begin{aligned}
\nabla\cdot(\mathbf{E}+\mathbf{P}/\varepsilon_0) &= \rho_f/\varepsilon_0 \\
\nabla\times\mathbf{E} &= -\partial\mathbf{B}/\partial t \\
\nabla\cdot\mathbf{B} &= 0 \\
c^2\nabla\times\mathbf{B} &= \mathbf{J}_f/\varepsilon_0+\frac{\partial}{\partial t}(\mathbf{E}+\mathbf{P}/\varepsilon_0)
\end{aligned}$$

where $\rho_f$ and $\mathbf{J}_f$ are the free charge density and free current density respectively. Assuming that there are no charges and currents present other than those due to polarization of the medium, the Maxwell equations can be written as

$$\nabla\cdot(\mathbf{E}+\mathbf{P}/\varepsilon_0) = 0 \tag{6.1}$$

$$\nabla\times\mathbf{E} = -\partial\mathbf{B}/\partial t \tag{6.2}$$

$$\nabla\cdot\mathbf{B} = 0 \tag{6.3}$$

$$c^2\nabla\times\mathbf{B} = \frac{\partial}{\partial t}(\mathbf{E}+\mathbf{P}/\varepsilon_0) \tag{6.4}$$

We now write the equations fully and examine them for changes which occur on crossing the boundary.

Equation (6.1) is

$$\frac{\partial E_x}{\partial x}+\frac{\partial E_y}{\partial y}+\frac{\partial E_z}{\partial z}+\frac{1}{\varepsilon_0}\frac{\partial P_x}{\partial x}+\frac{1}{\varepsilon_0}\frac{\partial P_y}{\partial y}+\frac{1}{\varepsilon_0}\frac{\partial P_z}{\partial z} = 0 \tag{6.5}$$

We are looking for changes which are associated with the boundary and we have already said that the only way in which the boundary can exert an influence is through a variation in $x$. With this in mind the $y$ and $z$ derivatives can be set to zero (changing with $y$ or $z$ will not find the boundary) and Eqn (6.5) becomes

$$\frac{\partial E_x}{\partial x}+\frac{1}{\varepsilon_0}\frac{\partial P_x}{\partial x} = 0 \tag{6.6}$$

that is

$$\frac{\partial}{\partial x}(E_x+P_x/\varepsilon_0) = 0 \tag{6.7}$$

Equation (6.7) can be interpreted as stating that the term $(E_x+P_x/\varepsilon_0)$ does not change in crossing the boundary, that is, $(E_x+P_x/\varepsilon_0)$ has the same value in medium 1 and medium 2 on either side of the boundary

$$(E_x+P_x/\varepsilon_0)_1 = (E_x+P_x/\varepsilon_0)_2 \tag{6.8}$$

This interpretation of Eqn (6.7) relies on our previous discussion of the boundary which we are regarding as a region of continuous change from medium 1 to medium 2 and consequently the change in any parameter on crossing the boundary must be a continuous function of $x$. If the $x$-derivative of a function is continuous and equal to zero then the function does not vary with $x$.

We can simplify the boundary condition on the $x$-components of the electric fields by making use of the relationship between **E** and **P** for an isotropic dielectric medium. In Chapter 4 we defined the relative permittivity $\varepsilon$ by

$$\varepsilon = (E+P/\varepsilon_0)/E$$

and it was shown that the permittivity was related to the refractive index $n$ by

$$\varepsilon = n^2$$

Therefore we can write

$$n^2 = (E+P/\varepsilon_0)/E$$

which gives the relationship

$$P = \varepsilon_0 E(n^2-1) \tag{6.9}$$

We can now write the boundary condition on the $x$-component of the electric fields as

$$[E_x+E_x(n_1^2-1)]_1 = [E_x+E_x(n_2^2-1)]_2$$

that is

$$n_1^2(E_x)_1 = n_2^2(E_x)_2 \tag{6.10}$$

Now consider the second Maxwell equation

$$\nabla\times\mathbf{E} = -\partial\mathbf{B}/\partial t$$

In the boundary region it is only the derivatives with respect to $x$ that are of interest for our present purposes and so we can set the right hand side of this equation to zero. Writing out the left hand side in full gives

$$\nabla \times \mathbf{E} = \begin{vmatrix} \mathbf{i} & \mathbf{j} & \mathbf{k} \\ \dfrac{\partial}{\partial x} & \dfrac{\partial}{\partial y} & \dfrac{\partial}{\partial z} \\ E_x & E_y & E_z \end{vmatrix}$$

and again all derivatives other than those with respect to $x$ can be set to zero, to give

$$\nabla \times \mathbf{E} = \mathbf{i}0 - \mathbf{j}\, \partial E_z/\partial x + \mathbf{k}\, \partial E_y/\partial x \tag{6.11}$$

But in the boundary region we have written

$$\nabla \times \mathbf{E} = -\partial \mathbf{B}/\partial t = 0$$

that is

$$0 = \mathbf{i}0 - \mathbf{j}\, \partial E_z/\partial x + \mathbf{k}\, \partial E_y/\partial x \tag{6.12}$$

and since this is a vector equation the components are separately equal to zero, which gives

$$\frac{\partial E_z}{\partial x} = 0 \tag{6.13}$$

and

$$\frac{\partial E_y}{\partial x} = 0 \tag{6.14}$$

From the previous discussion, Eqns (6.13) and (6.14) can be interpreted as stating that neither the $z$-component nor the $y$-component of the electric field changes on crossing the boundary. So the boundary conditions on the electric field are

$$\boxed{\begin{aligned} n_1^2(E_x)_1 &= n_2^2(E_x)_2 \\ (E_y)_1 &= (E_y)_2 \\ (E_z)_1 &= (E_z)_2 \end{aligned}} \tag{6.15}$$

The two remaining Maxwell equations,

$$\nabla \cdot \mathbf{B} = 0 \quad \text{and} \quad c^2 \nabla \times \mathbf{B} = \frac{\partial}{\partial t}(\mathbf{E} + \mathbf{P}/\varepsilon_0)$$

yield the boundary conditions for the magnetic field as

$$\boxed{\mathbf{B}_1 = \mathbf{B}_2} \tag{6.16}$$

which states that none of the components of the magnetic field change in crossing the boundary. The derivation of this result is left as an exercise: see Appendix 6.1 for the detailed working.

## 6.2 THE PROPAGATION VECTOR

In Chapter 4 it was shown that an equation of the form

$$\mathbf{E} = \mathbf{E}_0 \exp[i(\omega t - kz)] \tag{6.17}$$

describes a plane monochromatic wave travelling in the $z$-direction, where $\mathbf{E}_0$ is a constant vector whose magnitude determines the amplitude of the electric field and whose direction gives the direction of polarization of the electric field. The constant $k$ is given by

$$k = \omega n / c \tag{6.18}$$

where $c$ is the velocity of an electromagnetic wave in free space, $n$ is the refractive index of the medium in which the wave is travelling and $\omega$ is the angular frequency of the wave. We now need greater flexibility in specifying the direction of propagation of the wave. Consider the scalar product of the position vector $\mathbf{r}$ (the vector describing the field point) and a vector $\mathbf{K}$

$$\mathbf{K}\cdot\mathbf{r} = k_x x + k_y y + k_z z \tag{6.19}$$

The last term on the right hand side is similar to the spatial term $kz$ in the exponential for the unidirectional plane monochromatic wave described by Eqn (6.17) above. If we now define $\mathbf{K}$ as our propagation vector and write the equation for the plane monochromatic wave as

$$\mathbf{E} = \mathbf{E}_0 \exp[i(\omega t - \mathbf{K}\cdot\mathbf{r})]$$

the direction of propagation of the wave can be changed by simply changing the components of $\mathbf{K}$. The $z$-directional case can be obtained by setting $k_x = 0$ and $k_y = 0$ when the wave propagates in the $z$-direction only.

The description of the wave in terms of the propagation vector $\mathbf{K}$ has an additional benefit in that it allows the time and space derivatives to be obtained very easily. Writing out the scalar product in Eqn (6.20) gives

$$\mathbf{E} = \mathbf{E}_0 \exp[i(\omega t - k_x x - k_y y - k_z z)] \tag{6.21}$$

from which it is clear that

$$\frac{\partial \mathbf{E}}{\partial t} = i\omega \mathbf{E}$$

and consequently the operation $\partial/\partial t$ is equivalent to multiplying by $i\omega$

$$\boxed{\frac{\partial}{\partial t} \equiv i\omega} \tag{6.22}$$

The space derivatives are

$$\frac{\partial \mathbf{E}}{\partial x} = -ik_x\mathbf{E}, \qquad \frac{\partial \mathbf{E}}{\partial y} = -ik_y\mathbf{E}, \qquad \frac{\partial \mathbf{E}}{\partial z} = -ik_z\mathbf{E}$$

and since

$$\nabla \equiv \mathbf{i}\,\partial/\partial x+\mathbf{j}\,\partial/\partial y+\mathbf{k}\,\partial/\partial z$$

it follows that the operation $\nabla$ is equivalent to $-i\mathbf{K}$

$$\boxed{\nabla \equiv -i\mathbf{K}} \tag{6.23}$$

So the derivative operators $\partial/\partial t$ and $\nabla$ can be effected by multiplication.

In addition

$$\mathbf{K}\cdot\mathbf{K} = K^2$$

but Eqn (6.18) states that $k = \omega n/c$ and we have seen that the $z$-directional case is a particular case of the more general $\mathbf{k}$ and so we can write

$$\boxed{\mathbf{K}\cdot\mathbf{K} = K^2 = k_x^2+k_y^2+k_z^2 = \omega^2 n^2/c^2} \tag{6.24}$$

## 6.3 REFLECTION AND REFRACTION AT A PLANE BOUNDARY

The problem of reflection and refraction of a plane monochromatic wave at the plane boundary between two ideal dielectric media (assumed to be homogenous, isotropic and lossless) can now be restated in terms of the appropriate propagation vectors as illustrated in Fig. 6.2.

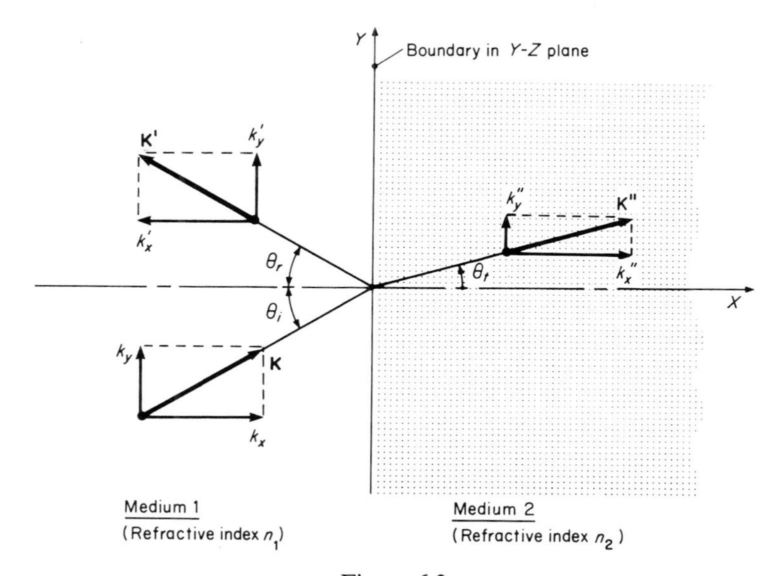


Figure 6.2

The equations for the electric fields of the waves can be written as:

the incident wave $\mathbf{E}_i = \mathbf{E}_0 \exp[i(\omega t - \mathbf{K}\cdot\mathbf{r})]$

the reflected wave $\mathbf{E}_r = \mathbf{E}_0' \exp[i(\omega' t - \mathbf{K}'\cdot\mathbf{r})]$

the transmitted wave $\mathbf{E}_t = \mathbf{E}_0'' \exp[i(\omega'' t - \mathbf{K}''\cdot\mathbf{r})]$

Obviously the incident plane monochromatic electric wave must be specified completely. This means specifying the direction of propagation by **K** and the direction of the electric field by $\mathbf{E}_0$. The direction in which the wave is travelling is arbitrary and the solution to the problem is not restricted by taking **K** to lie in the *x*-*y* plane in which case **K** is described by

$$\mathbf{K} = \mathbf{i}k_x + \mathbf{j}k_y + \mathbf{k}0$$

The direction of the incident electric field is given by the constant vector $\mathbf{E}_0$. In Chapter 4 it was seen that there was no rotation of the direction of the fields as a plane linearly polarized electromagnetic wave travels through an isotropic dielectric medium. As medium 1, medium 2 and the boundary region are each assumed to be isotropic there is no reason to suggest that there will be rotation of the fields in this case. If we specify the direction of $\mathbf{E}_0$ for the incident wave, then $\mathbf{E}_0'$ for the reflected wave and $\mathbf{E}_0''$ for the transmitted wave will be in the same direction as $\mathbf{E}_0$. However, there are two extreme possibilities for the direction of $\mathbf{E}_0$. As we have arranged the problem, $\mathbf{E}_0$ can be taken to be either parallel to the *x*-*y* plane or perpendicular to the *x*-*y* plane; all other possible orientations of $\mathbf{E}_0$ (where $\mathbf{E}_0$ is, of course, still perpendicular to the direction of propagation **K**) can then be represented by the sum of these two mutually perpendicular directions. We consider the two extreme cases separately.

### Case I

### The Plane of Polarization of E Perpendicular to the X-Y Plane (the Plane of Incidence)

In this case the electric fields have only a *z*-component. The electric field $\mathbf{E}_1$, at the boundary on the side of medium 1, is given by the sum of the incident and reflected fields, $\mathbf{E}_i$ and $\mathbf{E}_r$. On the side of medium 2 of the boundary the electric field $\mathbf{E}_2$ is simply the transmitted field $\mathbf{E}_t$. Therefore

$$\mathbf{E}_1 = \mathbf{E}_i + \mathbf{E}_r$$

and

$$\mathbf{E}_2 = \mathbf{E}_t$$

Each of these fields has only a *z*-component and the boundary condition on the *z*-component of the electric field is

$$(E_z)_1 = (E_z)_2$$

Consequently

$$\mathbf{E}_i + \mathbf{E}_r = \mathbf{E}_t$$

that is

$$\mathbf{E}_0 \exp[i(\omega t-\mathbf{K}\cdot\mathbf{r})]+\mathbf{E}_0' \exp[i(\omega' t-\mathbf{K}'\cdot\mathbf{r})] = \mathbf{E}_0'' \exp[i(\omega'' t-\mathbf{K}''\cdot\mathbf{r})] \quad (6.25)$$

We specify the direction of propagation of the incident wave to be in the $x$-$y$ plane, that is

$$\mathbf{K}\cdot\mathbf{r} = k_x x+k_y y+0z$$

and for the reflected wave

$$\mathbf{K}'\cdot\mathbf{r} = k_x' x+k_y' y+k_z' z$$

and for the transmitted wave

$$\mathbf{K}''\cdot\mathbf{r} = k_x'' x+k_y'' y+k_z'' z$$

Equation (6.25) holds at the boundary, that is, at $x = 0$, and we can therefore write

$$E_0 \exp[i(\omega t-k_y y-0z)]+E_0' \exp[i(\omega' t-k_y' y-k_z' z)] = E_0'' \exp[i(\omega'' t-k_y'' y-k_z'' z)]$$

This equation must hold at all points on the boundary, that is, for all values of $y$ and $z$ and it must hold for all values of the time $t$.

The only way in which this can be so is if

$$\omega = \omega' = \omega''$$
$$k_y = k_y' = k_y''$$
$$0 = k_z' = k_z''$$

We can make use of the general result given in Eqn (6.24) that

$$K^2 = k_x^2+k_y^2+k_z^2 = \omega^2 n^2/c^2$$

For the incident wave

$$K^2 = k_x^2+k_y^2 = \omega^2 n_1^2/c^2$$

and for the reflected wave

$$(K')^2 = (k_x')^2+(k_y')^2 = (\omega')^2 n_1^2/c^2$$

that is, since $\omega = \omega'$ and $k_y = k_y'$,

$$(k_x')^2+(k_y')^2 = \omega^2 n_1^2/c^2$$

$$\therefore \qquad k_x^2 = (k_x')^2 \quad \text{or} \quad k_x = \pm k_x'$$

The positive root in this equation would mean that the incident wave and the reflected wave would have the same components in their propagation vectors, that is, the waves would be travelling in the same direction. Consequently we take

$$k_x = -k_x'$$

as appropriate in this case. This result, taken with $k_y = k_y'$ means (see Fig. 6.3)

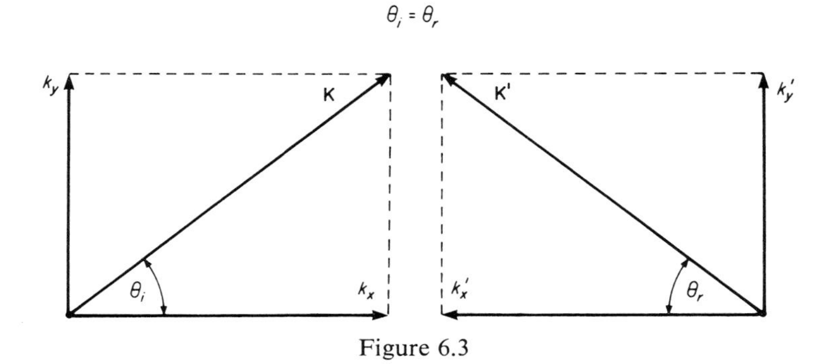


Figure 6.3

that the angle of incidence equals the angle of reflection

$$\boxed{\theta_i = \theta_r}$$

For the transmitted wave

$$(K'')^2 = (k_x'')^2 + (k_y'')^2 = (\omega'')^2 n_2^2/c^2$$

$$= \left(\frac{n_2}{n_1}\right)^2 (k_x^2 + k_y^2)$$

by comparison with the incident wave. But

$$\frac{k_y}{\sqrt{k_x^2 + k_y^2}} = \sin\theta_i$$

and

$$\frac{k_y''}{\sqrt{(k_x'')^2 + (k_y'')^2}} = \sin\theta_t$$

where

$$k_y = k_y''$$

$$\therefore \qquad (k_y'')^2/[(k_x'')^2 + (k_y'')^2] = k_y^2 \bigg/ \left[\left(\frac{n_2}{n_1}\right)^2 (k_x^2 + k_y^2)\right]$$

that is,

$$(\sin\theta_t)^2 = \left(\frac{n_1}{n_2}\right)^2 (\sin\theta_i)^2$$

or

$$\boxed{n_2 \sin\theta_t = n_1 \sin\theta_i}$$

which is Snell's law.

The results obtained so far show that in this case there is no change in frequency on reflection or refraction, all three waves travel parallel to the $x$-$y$ plane (the plane of incidence), the angle of incidence equals the angle of reflection, and that Snell's law relates the angle of incidence to the angle of transmission. To find the relationships between the amplitudes of the waves we turn to the magnetic boundary condition which is

$$\mathbf{B}_1 = \mathbf{B}_2$$

In this case the magnetic field $\mathbf{B}_1$ on the incident side of the boundary consists of the incident magnetic field $\mathbf{B}_i$ and the reflected magnetic field $\mathbf{B}_r$, while the field on the other side of the boundary is just $\mathbf{B}_t$. The boundary condition can be written as

$$\mathbf{B}_i + \mathbf{B}_r = \mathbf{B}_t$$

Each of these magnetic fields is related to the corresponding electric fields by the Maxwell equation

$$\nabla \times \mathbf{E} = -\partial \mathbf{B}/\partial t$$

In Section 6.2 it was shown that $\nabla \equiv -i\mathbf{K}$ and $\partial/\partial t \equiv i\omega$ and so this equation can be written as

$$\mathbf{B} = \frac{\mathbf{K} \times \mathbf{E}}{\omega}$$

Writing the magnetic boundary condition in terms of the electric fields gives

$$\frac{\mathbf{K} \times \mathbf{E}_i}{\omega} + \frac{\mathbf{K}' \times \mathbf{E}_r}{\omega'} = \frac{\mathbf{K}'' \times \mathbf{E}_t}{\omega''}$$

where we have seen that $\omega = \omega' = \omega''$. Each of the electric fields has a component in the $z$-direction only and we can therefore write

$$\begin{vmatrix} \mathbf{i} & \mathbf{j} & \mathbf{k} \\ k_x & k_y & 0 \\ 0 & 0 & E_{iz} \end{vmatrix} + \begin{vmatrix} \mathbf{i} & \mathbf{j} & \mathbf{k} \\ k_x' & k_y' & 0 \\ 0 & 0 & E_{rz} \end{vmatrix} = \begin{vmatrix} \mathbf{i} & \mathbf{j} & \mathbf{k} \\ k_x'' & k_y'' & 0 \\ 0 & 0 & E_{tz} \end{vmatrix} \qquad (6.26)$$

As a vector equation, the components must be separately equal. The $\mathbf{i}$ components give

$$k_y E_{iz} + k_y' E_{rz} = k_y'' E_{tz}$$

and since $k_y = k_y' = k_y''$ we can write this as

$$E_0 \exp[i(\omega t - k_y y)] + E_0' \exp[i(\omega' t - k_y' y)] = E_0'' \exp[i(\omega'' t - k_y'' y)]$$

where we are making use of our results that $0 = k_z' = k_z''$ and that $x = 0$ at the boundary. The exponential factor is the same in each term,

$$\therefore \qquad E_0 + E_0' = E_0''$$

Although this is an equation relating the amplitudes of the three electric fields, since all three electric fields are in the $z$-direction we can write

$$\mathbf{E}_0 + \mathbf{E}_0' = \mathbf{E}_0'' \tag{6.27}$$

Writing out the $\mathbf{j}$ components of Eqn (6.26) gives

$$-k_x E_{iz} - k_x' E_{rz} = -k_x'' E_{tz}$$

that is

$$k_x E_0 \exp[i(\omega t - k_y y)] + k_x' E_0' \exp[i(\omega' t - k_y' y)] = k_x'' E_0'' \exp[i(\omega'' t - k_y'' y)]$$

where we are again making use of the result that $0 = k_z' = k_z''$ and that $x = 0$ at the boundary. The exponential factor is the same in each term,

$$\therefore \qquad k_x E_0 + k_x' E_0' = k_x'' E_0''$$

and again since $\mathbf{E}_0$, $\mathbf{E}_0'$ and $\mathbf{E}_0''$ are all in the same direction we can write this scalar equation in terms of the vector quantities to give

$$k_x \mathbf{E}_0 + k_x' \mathbf{E}_0' = k_x'' \mathbf{E}_0'' \tag{6.28}$$

We now have two equations, Eqns (6.27) and (6.28), which relate the three electric fields. Multiplying Eqn (6.27) by $k_x''$ and subtracting the equations gives

$$(k_x'' - k_x)\mathbf{E}_0 + (k_x'' - k_x')E_0' = 0$$

that is

$$\mathbf{E}_0' = \frac{k_x - k_x''}{k_x'' - k_x'} \mathbf{E}_0$$

but $k_x' = -k_x$

$$\therefore \qquad \mathbf{E}_0' = \frac{k_x - k_x''}{k_x'' + k_x} \mathbf{E}_0$$

and since $k_y = k_y''$,

$$\mathbf{E}_0' = \frac{k_x/k_y - k_x''/k_y''}{k_x''/k_y'' + k_x/k_y} \mathbf{E}_0$$

$$= \frac{\cot\theta_i - \cot\theta_t}{\cot\theta_i + \cot\theta_t} \mathbf{E}_0$$

$$= \frac{\sin\theta_t \cos\theta_i - \cos\theta_t \sin\theta_i}{\sin\theta_t \cos\theta_i + \cos\theta_t \sin\theta_i} \mathbf{E}_0$$

$$\therefore \qquad \boxed{\mathbf{E}_0' = \frac{\sin(\theta_t - \theta_i)}{\sin(\theta_t + \theta_i)} \mathbf{E}_0} \tag{6.29}$$

Multiplying Eqn (6.27) by $k'_x$ and subtracting Eqn (6.28) gives

$$(k'_x - k_x)\mathbf{E}_0 = (k'_x - k''_x)\mathbf{E}''_0$$

that is

$$\mathbf{E}''_0 = \frac{2k_x}{k_x + k''_x}\,\mathbf{E}_0$$

because $k_x = -k'_x$. Dividing by $k_y = k''_y$ gives

$$\mathbf{E}''_0 = \frac{2k_x/k_y}{k_x/k_y + k''_x/k''_y}\,\mathbf{E}_0$$

$$= \frac{2\cot\theta_i}{\cot\theta_i + \cot\theta_t}\,\mathbf{E}_0$$

$$\boxed{\mathbf{E}''_0 = \frac{2}{1 + \tan\theta_i \cot\theta_t}\,\mathbf{E}_0} \qquad (6.30)$$

**Case II**

**The Plane of Polarization of E Parallel to the X-Y Plane (the Plane of Incidence)**

In this case the incident electric field will have an $x$ and $y$ component and can be described by

$$\mathbf{E}_i = \mathbf{E}_0 \exp[i(\omega t - \mathbf{K}\cdot\mathbf{r})]$$
$$= \mathbf{E}_0 \exp[i(\omega t - k_x x - k_y y - 0z)]$$

The incident wave is propagating in the $x$-$y$ plane and so has no $z$-component in its propagation vector $\mathbf{K}$. The constant vector $\mathbf{E}_0$ giving the direction of the incident field can be written as

$$\mathbf{E}_0 = \mathbf{i}E_{0x} + \mathbf{j}E_{0y} + \mathbf{k}0$$

We could continue to work with the electric fields as we did in Case I above, but in the present case it is better to work with the magnetic fields, as the interpretation of the results can be more easily made. The magnetic field in the incident wave can be described as

$$\mathbf{B}_i = \mathbf{B}_0 \exp[i(\omega t - \mathbf{K}\cdot\mathbf{r})]$$
$$= \mathbf{B}_0 \exp[i(\omega t - k_x x - k_y y - 0z)]$$

where the direction of the magnetic field is given by

$$\mathbf{B}_0 = \mathbf{i}0 + \mathbf{j}0 + \mathbf{k}B_0$$

and the magnetic field alternates with time and position along $\pm\mathbf{K}$.

Similarly for the reflected wave, we write

$$\mathbf{B}_r = \mathbf{B}'_0 \exp[i(\omega' t - \mathbf{K}'\cdot\mathbf{r})]$$
$$= \mathbf{B}'_0 \exp[i(\omega' t - k'_x x - k'_y y - k'_z z)]$$

where

$$\mathbf{B}_0' = \mathbf{i}0+\mathbf{j}0+\mathbf{k}B_0'$$

and for the transmitted wave

$$\begin{aligned}\mathbf{B}_t &= \mathbf{B}_0'' \exp[i(\omega''t-\mathbf{K}''\cdot\mathbf{r})] \\ &= \mathbf{B}_0'' \exp[i(\omega''t-k_x''x-k_y''y-k_z''z)]\end{aligned}$$

where

$$\mathbf{B}_0'' = \mathbf{i}0+\mathbf{j}0+\mathbf{k}B_0''$$

We can now apply the magnetic boundary condition, that is,

$$\mathbf{B}_1 = \mathbf{B}_2$$

The field on the side of medium 1 of the boundary consists of the sum of the incident and reflected waves whereas on the side of medium 2 of the boundary there is only the transmitted wave. As all regions are isotropic, there is no rotation of the field as it propagates and $\mathbf{B}_i$, $\mathbf{B}_r$ and $\mathbf{B}_t$ are all in the same direction: in this case they are in the $\mathbf{k}$ direction. The boundary condition can then be written as

$$\mathbf{B}_i+\mathbf{B}_r = \mathbf{B}_t$$

that is

$$\begin{aligned}B_0\mathbf{k}\exp[i(\omega t-k_xx-k_yy-0z)]+B_0'\mathbf{k}\exp[i(\omega't-k_x'x-k_y'y-k_z'z)] \\ = B_0''\mathbf{k}\exp[i(\omega''t-k_x''x-k_y''y-k_z''z)]\end{aligned}$$

As a boundary condition, this equation only holds at the boundary where $x = 0$, but it must hold at all points on the boundary, that is, for all $y$ and $z$, and it must hold at all times $t$. Consequently the following conditions apply

$$\begin{aligned}\omega &= \omega' = \omega'' \\ k_y &= k_y' = k_y'' \\ 0 &= k_z' = k_z''\end{aligned}$$

The last result shows that all three waves are propagated in the same plane (the plane of incidence) which is parallel to the $x$-$y$ plane. Taking the square of the magnitudes of the propagation vectors for incidence and reflection gives

$$\begin{aligned}K^2 &= k_x^2+k_y^2 = \omega^2n_1^2/c^2 \\ (K')^2 &= (k_x')^2+(k_y')^2 = (\omega')^2n_1^2/c^2\end{aligned}$$

and since $k_y' = k_y$ and $\omega = \omega'$, we have

$$k_x^2 = (k_x')^2$$

or

$$k_x = \pm k_x'$$

The positive root would mean that the reflected and incident waves were travelling in the same directions and consequently the negative root is appropriate here

$$\therefore \qquad k_x = -k'_x$$

Consequently

$$\theta_i = \theta_r$$

Considering now the incident and transmitted propagation vectors we have for the incident wave

$$K^2 = k_x^2 + k_y^2 = \omega^2 n_1^2 / c^2$$

and for the transmitted wave

$$(K'')^2 = (k''_x)^2 + (k''_y)^2 = (\omega'')^2 n_2^2 / c^2$$

but $k_y = k''_y$ and $\omega = \omega''$

$$\therefore \qquad k_y / \sqrt{k_x^2 + k_y^2} = \sin\theta_i = k_y \bigg/ \left(\frac{\omega n_1}{c}\right)$$

and

$$k''_y / \sqrt{(k''_x)^2 + (k''_y)^2} = \sin\theta_t = k''_y \bigg/ \left(\frac{\omega'' n_2}{c}\right)$$

$$\therefore \qquad \boxed{n_1 \sin\theta_i = n_2 \sin\theta_t}$$

which is Snell's law.

To obtain the relationships between the amplitudes of the waves we need to make use of the boundary conditions on the electric field. We can obtain the electric fields from the magnetic fields by means of the Maxwell equation

$$c^2 \nabla \times \mathbf{B} = \mathbf{J}_f / \varepsilon_0 + \frac{\partial}{\partial t} (\mathbf{E} + \mathbf{P} / \varepsilon_0)$$

where in this case we set the free current density to zero and write the equation for a homogeneous, isotropic, lossless dielectric medium as

$$c^2 \nabla \times \mathbf{B} = \frac{\partial}{\partial t} (\mathbf{E} + \mathbf{P} / \varepsilon_0)$$

The relative permittivity $\varepsilon$ has been defined as

$$\varepsilon = (E + P/\varepsilon_0)/E$$

and $\varepsilon$ is related to the refractive index $n$ by

$$n^2 = \varepsilon$$

Therefore we can write

$$n^2 = (E + P/\varepsilon_0)/E$$

and the Maxwell relation becomes

$$c^2\nabla\times\mathbf{B} = n^2\,\partial\mathbf{E}/\partial t$$

We have seen that $\partial/\partial t \equiv i\omega$ and $\nabla \equiv -i\mathbf{K}$ and so we obtain

$$\mathbf{E} = -\left(\frac{c^2}{\omega n^2}\right)[\mathbf{K}\times\mathbf{B}]$$

$$= \frac{c^2}{\omega n^2}\,\mathbf{B}\times\mathbf{K} \tag{6.31}$$

hence

$$\mathbf{E}_i = \left(\frac{c^2}{\omega n_1^2}\right)\begin{vmatrix} \mathbf{i} & \mathbf{j} & \mathbf{k} \\ 0 & 0 & B_i \\ k_x & k_y & 0 \end{vmatrix}$$

$$= -\left(\frac{c^2}{\omega n_1^2}\right)[\mathbf{i}k_yB_i - \mathbf{j}k_xB_i + \mathbf{k}0]$$

and similarly

$$\mathbf{E}_r = -\left(\frac{c^2}{\omega' n_1^2}\right)[\mathbf{i}k_y'B_r - \mathbf{j}k_x'B_r + \mathbf{k}0]$$

and

$$\mathbf{E}_t = -\left(\frac{c^2}{\omega'' n_2^2}\right)[\mathbf{i}k_y''B_t - \mathbf{j}k_x''B_t + \mathbf{k}0]$$

The boundary conditions for the electric fields are

$$(E_y)_1 = (E_y)_2$$

and

$$n_1^2(E_x)_1 = n_2^2(E_x)_2$$

Applying the $x$-boundary condition and noting that $\omega = \omega' = \omega''$ we obtain

$$\left(\frac{1}{n_1}\right)^2 n_1^2[k_yB_i + k_y'B_r] = n_2^2[k_y''B_t]\left(\frac{1}{n_2}\right)^2$$

and since $k_y = k_y' = k_y''$ we can write

$$B_i + B_r = B_t$$

or

$$B_0 + B_0' = B_0''$$

which can be written as a vector equation since all three magnetic fields are in the same direction, that is

$$\mathbf{B}_0 + \mathbf{B}_0' = \mathbf{B}_0'' \tag{6.32}$$

Applying the $y$-boundary condition gives

$$\left(\frac{1}{n_1}\right)^2 [k_x B_i + k_x' B_r] = \left(\frac{1}{n_2}\right)^2 k_x'' B_t$$

which, since $k_x = -k_x'$, can be written as

$$B_i - B_r = \left(\frac{n_1}{n_2}\right)^2 \left(\frac{k_x''}{k_x}\right)^2 B_t$$

or

$$B_0 - B_0' = \left(\frac{n_1}{n_2}\right)^2 \left(\frac{k_x''}{k_x}\right) B_0''$$

This equation can be written as a vector equation since all three magnetic fields are in the same direction

$$\mathbf{B}_0 - \mathbf{B}_0' = \left(\frac{n_1}{n_2}\right)^2 \left(\frac{k_x''}{k_x}\right) \mathbf{B}_0'' \qquad (6.33)$$

Multiplying Eqn (6.32) by $(n_1/n_2)^2 \times (k_x''/k_x)$ and subtracting Eqn (6.33) gives

$$\mathbf{B}_0 \left[\left(\frac{n_1}{n_2}\right)^2 \frac{k_x''}{k_x} - 1\right] + \mathbf{B}_0' \left[\left(\frac{n_1}{n_2}\right)^2 \frac{k_x''}{k_x} + 1\right] = 0$$

That is

$$\mathbf{B}_0' = -\mathbf{B}_0 \left[\frac{\left(\frac{n_1}{n_2}\right)^2 \frac{k_x''}{k_x} - 1}{\left(\frac{n_1}{n_2}\right)^2 \frac{k_x''}{k_x} + 1}\right]$$

But,

$$\left(\frac{n_1}{n_2}\right)^2 \frac{k_x''}{k_x} = \left(\frac{n_1}{n_2}\right)^2 \frac{k_x''}{k_y''} \frac{k_y}{k_x} \qquad \because\ k_y = k_y''$$

$$= \left(\frac{\sin\theta_t}{\sin\theta_i}\right)^2 \frac{\cos\theta_t}{\sin\theta_t} \frac{\sin\theta_i}{\cos\theta_i} \qquad \text{from Snell's law}$$

$$= \frac{\sin\theta_t \cos\theta_t}{\sin\theta_i \cos\theta_i}$$

$$\therefore \qquad \mathbf{B}_0' = -\mathbf{B}_0 \left(\frac{\sin\theta_t \cos\theta_t - \sin\theta_i \cos\theta_i}{\sin\theta_t \cos\theta_t + \sin\theta_i \cos\theta_i}\right)$$

or

$$\boxed{\mathbf{B}_0' = -\mathbf{B}_0 \frac{\tan(\theta_t - \theta_i)}{\tan(\theta_t + \theta_i)}} \qquad (6.34)$$

The relationship between the transmitted and incident magnetic fields is obtained by adding Eqns (6.32) and (6.33) to give

$$2\mathbf{B}_0 = \mathbf{B}_0''\left[1+\left(\frac{n_1}{n_2}\right)^2\frac{k_x''}{k_x}\right]$$

$$\mathbf{B}_0'' = \frac{2}{\left[1+\left(\frac{n_1}{n_2}\right)^2\frac{k_x''}{k_x}\right]}\mathbf{B}_0$$

or

$$\boxed{\mathbf{B}_0'' = \frac{2}{\left(1+\frac{n_1}{n_2}\frac{\cos\theta_t}{\cos\theta_i}\right)}\mathbf{B}_0} \qquad (6.35)$$

Our results relating the magnetic fields can now be used to give the corresponding relationships between the *amplitudes* of the electric fields. Equation (6.31) above gives the relationship between the electric and magnetic fields as

$$\mathbf{E} = \frac{c^2}{\omega n^2}\mathbf{B}\times\mathbf{K}$$

where $K = \omega n/c$. Therefore we can write

$$E = \frac{c^2}{\omega n^2}B\left(\frac{\omega n}{c}\right)$$

$$E = \frac{c}{n}B$$

Then from Eqn (6.34) we can relate the amplitudes of the electric fields in the incident and reflected waves

$$\frac{E_0'}{E_0} = \frac{\tan(\theta_t-\theta_i)}{\tan(\theta_t+\theta_i)} \qquad (6.36)$$

where the ratio of the magnitudes can only be taken as positive.

From Eqn (6.35) we obtain

$$n_2E_0'' = \frac{2}{\left(1+\frac{n_1}{n_2}\frac{\cos\theta_t}{\cos\theta_i}\right)}n_1E_0$$

$$\frac{E_0''}{E_0} = \left(\frac{n_1}{n_2}\right)\left[\frac{2}{1+\frac{n_1}{n_2}\frac{\cos\theta_t}{\cos\theta_i}}\right] \qquad (6.37)$$

## 6.4 POLARIZATION BY REFLECTION

The amplitudes of the electric field in the reflected wave is given by Eqns (6.29) and (6.36), that is,

$$\text{for } \mathbf{E} \perp \text{ plane of incidence} \quad \frac{E_0'}{E_0} = \frac{\sin(\theta_t - \theta_i)}{\sin(\theta_t + \theta_i)} \qquad (6.29)$$

$$\text{for } \mathbf{E} \parallel \text{ plane of incidence} \quad \frac{E_0'}{E_0} = \frac{\tan(\theta_t - \theta_i)}{\tan(\theta_t + \theta_i)} \qquad (6.36)$$

However, $\tan(\pi/2) = \infty$ and if $\theta_t + \theta_i = \pi/2$ then Eqn (6.36) shows that there is no reflected component parallel to the $x$-$y$ plane. There is still a reflected component perpendicular to the $x$-$y$ plane since $\sin(\pi/2) = 1$ and Eqn (6.29) shows that the ratio $E_0'/E_0$ is finite. Another way in which Eqn (6.36) and Eqn (6.29) could be zero is if $\theta_t = \theta_i \neq 0$ when the numerator in both expressions is zero, but $\theta_t = \theta_i$ cannot physically correspond to reflection since Snell's law shows that if $\theta_t = \theta_i$ (and is non-zero) then $n_2 = n_1$ and there is no boundary. We conclude from Eqns (6.29) and (6.36) that if $\theta_t + \theta_i = \pi/2$, then there is polarization by reflection from the boundary since only the component of the electric field which is perpendicular to the plane of incidence will be reflected.

In general there will be some polarization for any angle of incidence but there is complete polarization by reflection if

$$\boxed{\theta_t + \theta_i = \pi/2} \qquad (6.38)$$

and when this equation is taken with Snell's law, it defines a particular angle of incidence, the Brewster Angle, $\theta_B$ for which there is complete polarization on reflection at the boundary between the given media. It follows that $\tan\theta_B = n_2/n_1$.

## 6.5 NORMAL INCIDENCE

In this case the ratio of the reflected amplitude $E_0'$ to the incident amplitude $E_0$ takes a particularly simple form. From Snell's law it follows that for normal incidence

$$\theta_i = 0 = \theta_t$$

As the ratio 0/0 is indeterminate we must change the form of Eqns (6.29) and (6.36); we had for $\mathbf{E}$ parallel to the plane of incidence

$$\frac{E_0'}{E_0} = \frac{\tan(\theta_t - \theta_i)}{\tan(\theta_t + \theta_i)} = \frac{\cos\theta_t \sin\theta_t - \cos\theta_i \sin\theta_i}{\cos\theta_t \sin\theta_t + \cos\theta_i \sin\theta_i}$$

$$= \frac{\cos\theta_t - \cos\theta_i(\sin\theta_i/\sin\theta_t)}{\cos\theta_t + \cos\theta_i(\sin\theta_i/\sin\theta_t)}$$

and if $\theta_t = \theta_i = 0$, making use of Snell's law we obtain

$$\frac{E_0'}{E_0} = \frac{n_1 - n_2}{n_1 + n_2}$$

Similarly, for **E** perpendicular to the plane of incidence, we had

$$\frac{E_0'}{E_0} = \frac{\sin(\theta_t - \theta_i)}{\sin(\theta_t + \theta_i)} = \frac{\sin\theta_t \cos\theta_i - \cos\theta_t \sin\theta_i}{\sin\theta_t \cos\theta_i + \cos\theta_t \sin\theta_i}$$

and again from Snell's law, if $\theta_i = \theta_t = 0$, we obtain $E_0'/E_0 = (n_1 - n_2)/(n_1 + n_2)$. So for normal incidence, the results for **E** perpendicular to the plane of incidence and those for **E** parallel to the plane of incidence are identical, as would be expected. The intensity or energy of the wave is proportional to the square of the amplitude of either the **E** or **B** field ($\mathbf{S} = \varepsilon_0 c^2 \mathbf{E} \times \mathbf{B}$ and $E = (c/n)B$), and the ratio of the reflected intensity $I_r$ to the incident intensity $I_i$ is given by

$$\frac{I_r}{I_i} = \left(\frac{E_0'}{E_0}\right)^2 = \left(\frac{n_1 - n_2}{n_1 + n_2}\right)^2$$

In the case of normal incidence, the reflection at a boundary is a function of the square of the difference in the refractive indices.

## 6.6 PHASE CHANGE ON REFLECTION

Equations (6.29) and (6.34) relate the incident and reflected fields. These equations are

for $\mathbf{E}_\perp$
$$\mathbf{E}_0' = \frac{\sin(\theta_t - \theta_i)}{\sin(\theta_t + \theta_i)} \mathbf{E}_0 \qquad (6.29)$$

for $\mathbf{E}_\parallel$
$$\mathbf{B}_0' = -\frac{\tan(\theta_t - \theta_i)}{\tan(\theta_t + \theta_i)} \mathbf{B}_0 \qquad (6.34)$$

In the first case we can see that Eqn (6.29) says that $\mathbf{E}_0'$ and $\mathbf{E}_0$ will be in the same direction provided the ratio of the sine terms is positive. The term $\sin(\theta_t + \theta_i)$ is always positive since the angle $(\theta_t + \theta_i)$ must lie between 0 and $\pi$ radians. However, the term $\sin(\theta_t - \theta_i)$ will be *negative* if $\theta_i > \theta_t$ since then the angle $(\theta_t - \theta_i)$ will be between 0 and $-\pi/2$. From Snell's law the condition $\theta_i > \theta_t$ can be seen to be equivalent to the condition $n_2 > n_1$.

So if the incident wave is reflected from a medium of greater refractive index, then Eqn (6.29) states that, for the component of the electric field perpendicular to the plane of incidence, the incident and reflected fields will be in opposite directions at the boundary. That is, if $n_2 > n_1$, then the perpendicular component of the electric field undergoes a phase change of 180° (see Fig. 6.4).

But what about the second case where the electric field is in the plane of incidence? The relationship between the magnetic fields is given by Eqn (6.34) and

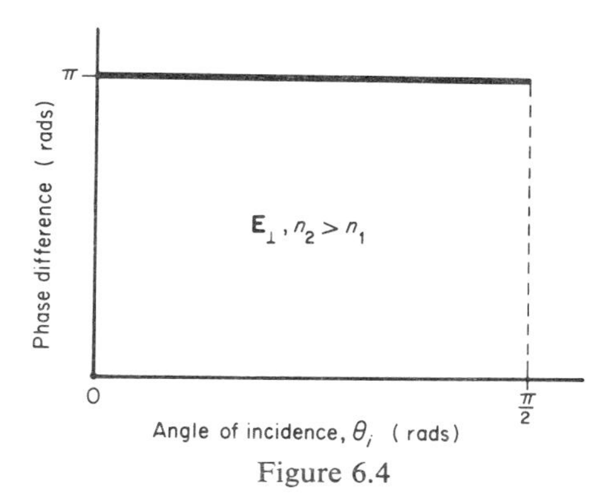


Figure 6.4

if there is to be a phase change in the magnetic field then the sign of $-\tan(\theta_t-\theta_i)/\tan(\theta_t+\theta_i)$ must be negative. For $(\theta_t+\theta_i) < \pi/2$, $\tan(\theta_t+\theta_i)$ is positive and if we again consider $\theta_i > \theta_t$ then $-\tan(\theta_t-\theta_i)$ is positive. So for $n_2 > n_1$ and $(\theta_t+\theta_i) < \pi/2$ there is *no* phase change in the magnetic field. If we now consider angles of incidence greater than the Brewster angle so that $(\theta_t+\theta_i) > \pi/2$, then $\tan(\theta_t+\theta_i)$ is negative and for such angles we do have a phase change of 180° between the incident and reflected magnetic fields in this case. Our results are illustrated in Fig. 6.5.

The sudden change in phase at the Brewster angle does not correspond to a dramatic physical switching at this particular angle of incidence since, as we have shown above, the amplitude of the reflected wave is zero at the Brewster angle for the case of the electric field in the plane of incidence.

But we must be careful now not to say that our result for the magnetic field is the same as that for the electric field. To obtain the electric field from the magnetic field we must use Eqn (6.31) above, that is

$$\mathbf{E} = \frac{c^2}{\omega n^2}\,\mathbf{B}\times\mathbf{K} \tag{6.31}$$

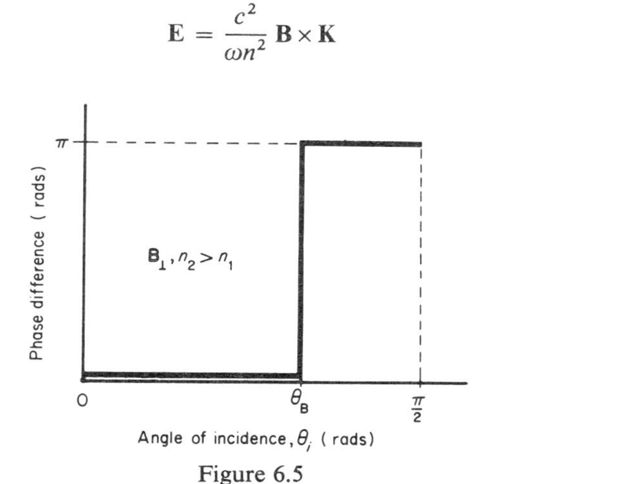


Figure 6.5

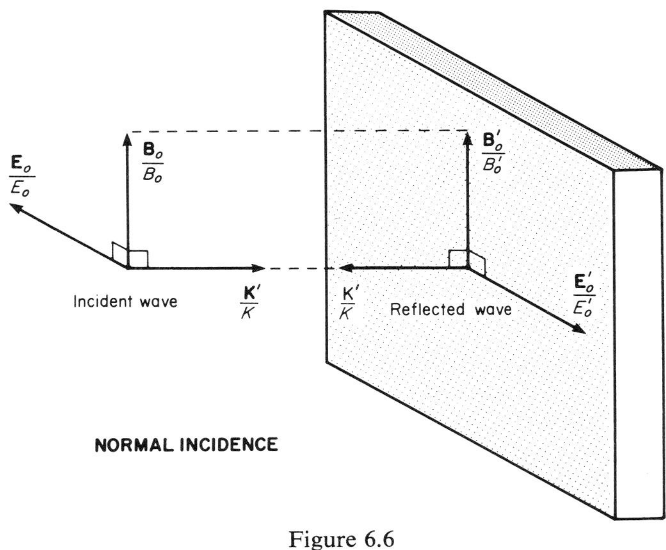


Figure 6.6

The direction of the electric field does not depend only on **B**, but in addition we need to consider the direction of propagation as given by **k**. The direction of propagation for the incident and reflected waves is, of course, not the same. We can only meaningfully discuss the relative phases of the incident and reflected waves in this case for the two limiting values for the angle of incidence (a) $\theta_i \to 0$, (b) $\theta_i \to \pi/2$. As $\theta_i \to 0$ then the incident and reflected directions of propagation

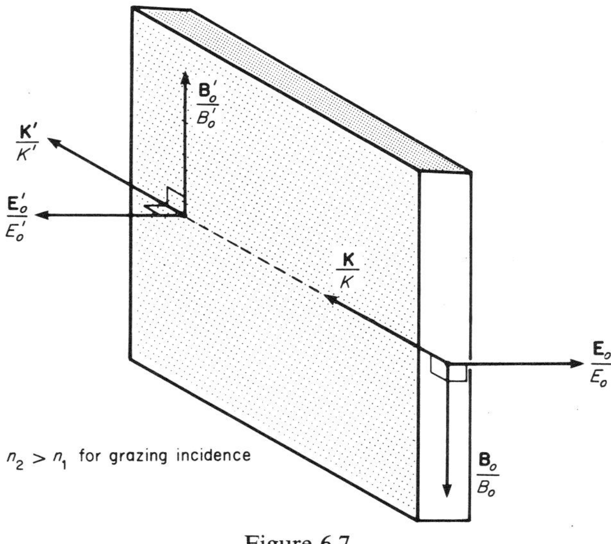


Figure 6.7

become opposed (see Fig. 6.6) so that

$$\mathbf{K} = -\mathbf{K}'$$

and we see that

$$\frac{\mathbf{E}_0'}{E_0'} = -\frac{\mathbf{E}_0}{E_0}$$

which corresponds to a phase change of 180° for $\theta_i > \theta_t$ that is, $n_2 > n_1$. So for normal incidence, if $n_2 > n_1$, then the electric field changes phase by 180° on reflection, regardless of its orientation to the plane of incidence—this is as it should be since for normal incidence there can be no distinction between Case I and Case II. At the other extreme, grazing incidence, the directions of propagation of the incident and reflected waves approach each other (see Fig. 6.7) as $\theta_i \to \pi/2$ and we write

$$\mathbf{K} = \mathbf{K}'$$

and so the two electric fields have the same phase relationship as the two magnetic fields

$$\frac{\mathbf{E}_0'}{E_0'} = -\frac{\mathbf{E}_0}{E_0}$$

and consequently the incident and reflected electric fields are now out of phase by 180° as $\theta_i \to \pi/2$.

For the condition $n_2 < n_1$, then the phase relations are reversed and we can summarize the results as indicated in Figs 6.8 and 6.9.

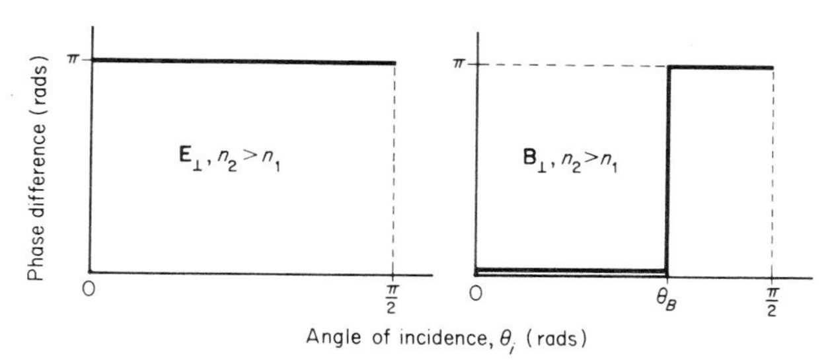


Figure 6.8

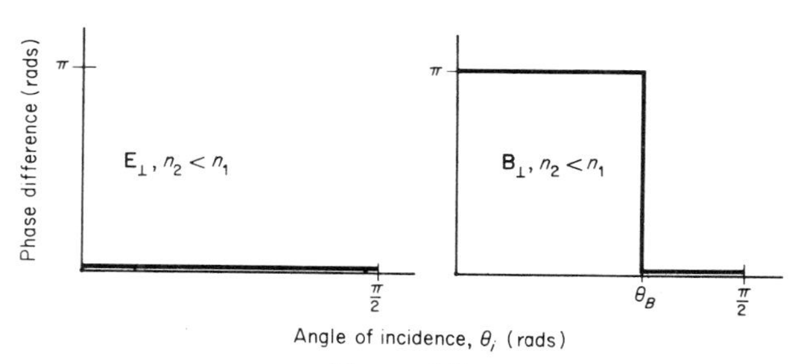


Figure 6.9

The phase relations for the $\mathbf{B}_{\parallel}$ and $\mathbf{E}_{\parallel}$ fields are obtained from $\mathbf{E}_{\perp}$ and $\mathbf{B}_{\perp}$ using

$$\mathbf{B} = \left(\frac{1}{\omega}\right) \mathbf{K} \times \mathbf{E}$$

and

$$\mathbf{E} = \left(\frac{c^2}{\omega n^2}\right) \mathbf{B} \times \mathbf{K}$$

## 6.7 REAL MATERIALS

For simplicity we have considered isotropic materials, that is, the somewhat restricted range of materials where the properties are the same in all directions, and we have considered the refractive index to be entirely real, that is, the materials are 'lossless'. In addition we have taken the materials to be homogeneous: they are perfectly uniform with no structural defects or impurities. In general, real materials are not homogeneous and the defects behave as small boundaries which may give rise to reflections within the material. An example of such a defect would be a small void in a solid material, where on crossing the void the refractive index could drop from that of the material to, say, that of vacuum and then increase sharply again to that of the material. The void introduces differences in refractive index, which, as we have seen above, can give rise to reflection. Consequently, if materials are to be very transparent they must be as nearly homogeneous as possible. Particle free gases and liquids are transparent apart from the specific regions where they absorb as discussed in Section 4.6. In the visible region of the electromagnetic spectrum, such materials may appear coloured due to such absorption of a particular frequency band, but otherwise the material will be transparent. We shall return to the topic of reflection from within a medium when we consider the scattering of electromagnetic radiation in Chapter 8.

## 6.8 SUMMARY OF RESULTS

1. There is no change in frequency on reflection or transmission at the boundary.

2. The incident, reflected and transmitted waves travel in the same plane, the plane of incidence.

3. The angle of incidence $\theta_i$ equals the angle of reflection $\theta_r$.

4. The angle of incidence $\theta_i$ in the medium of refractive index $n_1$ is related to the angle of transmission $\theta_t$ in the medium of refractive index $n_2$ by Snell's law

$$n_1 \sin\theta_i = n_2 \sin\theta_t$$

5. For the case where **E** is parallel to the plane of incidence the ratio of the amplitude $E_0'$ of the electric field in the reflected wave to the amplitude $E_0$ of the electric field in the incident wave is given by

$$\frac{E_0'}{E_0} = \frac{\tan(\theta_t - \theta_i)}{\tan(\theta_t + \theta_i)}$$

and for the case where **E** is perpendicular to the plane of incidence

$$\frac{E_0'}{E_0} = \frac{\sin(\theta_t - \theta_i)}{\sin(\theta_t + \theta_i)}$$

6. For the case where **E** is parallel to the plane of incidence the ratio of the amplitude $E_0''$ of the electric field in the transmitted wave to the amplitude $E_0$ of the electric field in the incident wave is given by

$$\frac{E_0''}{E_0} = 2\left(\frac{n_1}{n_2}\right) \Big/ \left(1 + \frac{n_1 \cos\theta_t}{n_2 \cos\theta_i}\right)$$

and for the case where **E** is perpendicular to the plane of incidence

$$\frac{E_0''}{E_0} = \frac{2}{1 + \tan\theta_i \cot\theta_t}$$

7. There is complete polarization on reflection if $\theta_t + \theta_i = \pi/2$ in which case the angle of incidence is known as the Brewster angle, $\theta_B$. For this angle of incidence the only component in the electric field of the reflected wave is perpendicular to the plane of incidence.

8. In the case of normal incidence, the ratio of the reflected and incident intensities is given by

$$\frac{I_r}{I_i} = \left(\frac{E_0'}{E_0}\right)^2 = \left(\frac{n_1 - n_2}{n_1 + n_2}\right)^2$$

9. The phase change on reflection is determined by the following equations;

for Case I, $\mathbf{E}_\perp$:

$$\mathbf{E}_0' = \frac{\sin(\theta_t - \theta_i)}{\sin(\theta_t + \theta_i)} \mathbf{E}_0$$

$$\mathbf{B} = \left(\frac{1}{\omega}\right) \mathbf{K} \times \mathbf{E}$$

which is the result for the component of the electric field which is perpendicular to the plane of incidence;

for Case II, $\mathbf{E}_\parallel$:

$$\mathbf{B}_0' = -\frac{\tan(\theta_t - \theta_i)}{\tan(\theta_t + \theta_i)} \mathbf{B}_0$$

$$\mathbf{E} = \left(\frac{c^2}{\omega n^2}\right) \mathbf{B} \times \mathbf{K}$$

which is the result for the component of the magnetic field which is perpendicular to the plane of incidence. The phase relationships can be summarized as shown in Fig. 6.10.

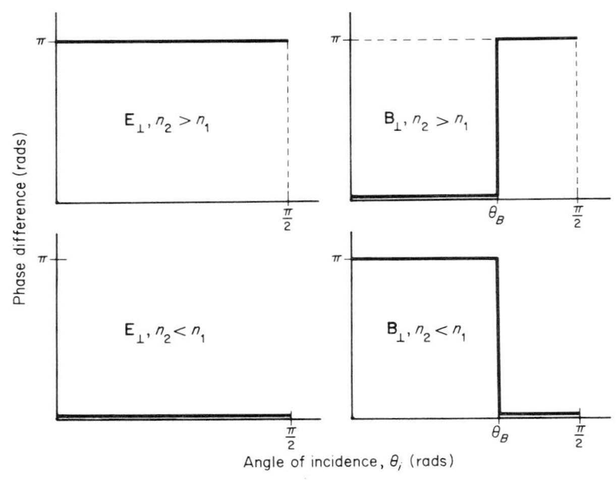


Figure 6.10

# APPENDIX 6.1

**Exercise: To Establish the Magnetic Boundary Conditions**

*Solution*

Equation (6.3) is

$$\nabla \cdot \mathbf{B} = 0$$

that is

$$\frac{\partial B_x}{\partial x} + \frac{\partial B_y}{\partial y} + \frac{\partial B_z}{\partial z} = 0$$

Setting all variations other than those with $x$ to zero gives

$$\frac{\partial B_x}{\partial x} = 0$$

If the derivative is continuous across the boundary this equation means that the $x$-component of the magnetic field does not change in crossing the boundary

$$(B_x)_1 = (B_x)_2$$

Equation (6.4) is

$$c^2 \nabla \times \mathbf{B} = \frac{\partial}{\partial t}(\mathbf{E} + \mathbf{P}/\varepsilon_0)$$

Only variations with $x$ are of interest in establishing the boundary conditions and variations with respect to $y$, $z$ and $t$ can be set to zero for the present purpose. Since

$$\nabla \times \mathbf{B} = \begin{vmatrix} \mathbf{i} & \mathbf{j} & \mathbf{k} \\ \dfrac{\partial}{\partial x} & \dfrac{\partial}{\partial y} & \dfrac{\partial}{\partial z} \\ B_x & B_y & B_z \end{vmatrix}$$

Eqn (6.4) reduces to

$$\mathbf{i}\left(\frac{\partial B_z}{\partial y} - \frac{\partial B_y}{\partial z}\right) - \mathbf{j}\left(\frac{\partial B_z}{\partial x} - \frac{\partial B_x}{\partial z}\right) + \mathbf{k}\left(\frac{\partial B_y}{\partial x} - \frac{\partial B_x}{\partial y}\right) = 0$$

that is

$$\mathbf{i}0 - \mathbf{j}\frac{\partial B_z}{\partial x} + \mathbf{k}\frac{\partial B_y}{\partial x} = 0$$

$$\therefore \qquad \frac{\partial B_z}{\partial x} = 0 \quad \text{and} \quad \frac{\partial B_y}{\partial x} = 0$$

Again, if the derivatives are continuous across the boundary, then

$$(B_z)_1 = (B_z)_2$$

and

$$(B_y)_1 = (B_y)_2$$

and the boundary conditions for the magnetic field can be summarized in the vector equation

$$\mathbf{B}_1 = \mathbf{B}_2$$

# PROBLEMS

1.
Show that the Brewster angle $\theta_B$ is given by:

$$\theta_B = \tan^{-1}(n_2/n_1)$$

Hence calculate the angle of incidence which gives complete polarization on reflection for visible light at (a) the air/glass interface, (b) the air/water interface. Take the approximate refractive indices of air, water and glass as 1.00, 1.33 and 1.52 respectively and take the reflection to be from the medium of higher refractive index.

2.
Show that for normal incidence the ratio of the amplitudes of the electric fields in the transmitted and incident waves becomes $E_0''/E_0 = 2/[1+(n_2/n_1)]$ for both Eqns (6.30) and (6.37).

3.
Special materials are required to transmit radiation over a wide frequency range in the infrared region of the electromagnetic spectrum. Two such materials which are commonly used are potassium bromide (refractive index 1.54 at $\lambda = 5\ \mu m$) and silver chloride (refractive index 2.00 at $\lambda = 5\ \mu m$). Estimate the percentage power loss at normal incidence in a beam of infrared radiation as it passes through plane parallel windows of these materials. Assume that the windows are in air (refractive index 1.00) and that the only energy losses occur on reflection at the window surfaces. Hint: If you are working in terms of the reflected waves be careful to consider the phase change at each interface and the boundary conditions on the electric field.

4.
Show that if we take $\theta_t = \pi/2$ in Eqns (6.29) and (6.34), then the incident and reflected waves have the same amplitude. Under this condition all the energy is being 'totally internally reflected'. Use Snell's law to show that total internal reflection will occur for all angles of incidence greater than a critical angle, $\theta_c$, where $\theta_c = \sin^{-1}(n_1/n_2)$. Hence calculate the critical angle for total internal reflection for visible light at a glass/air interface. Take the refractive indices for glass and air as 1.52 and 1.00 respectively.

5.
Many optical instruments make use of the fact that the critical angle for total internal reflection in common glasses is always less than 45°. Show that for visible light passing through the prism shown in Fig. 6.11 and being totally internally reflected there is only a small loss in energy. Take the prism to be in air (refractive index 1.00) and take the approximate refractive index of glass to be 1.52 for visible light. Hence estimate the percentage energy loss.

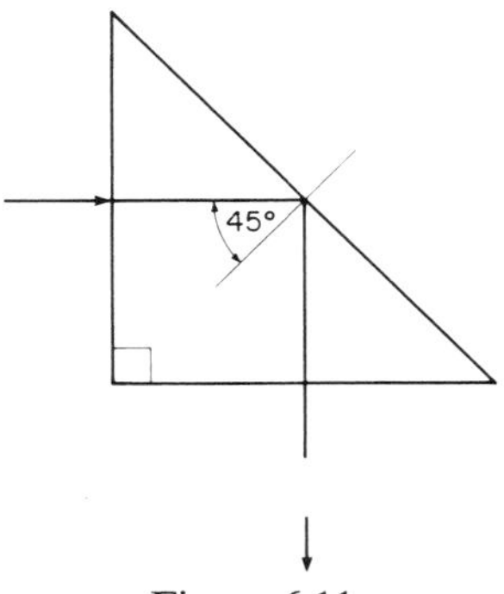


Figure 6.11

## FURTHER READING

For further information see Tenquist et al., Andrews, Paris & Hurd, Jenkins & White.

Chapter 7

# THE SCALAR POTENTIAL AND THE VECTOR POTENTIAL

Instead of dealing with the **E** and **B** fields, it is sometimes more convenient to deal with a potential function, that is, a function whose derivative will give the field. We now set out to find such functions for the **E** and **B** fields. The results we obtain in this chapter will form the basis of our treatment of radiating systems in Chapter 8.

## 7.1 THE VECTOR POTENTIAL A

Consider our third Maxwell equation which we have written as

$$\nabla \cdot \mathbf{B} = 0 \tag{7.1}$$

For *any* vector **C**

$$\nabla \cdot (\nabla \times \mathbf{C}) = 0 \tag{7.2}$$

Because the vector $\nabla \times \mathbf{C}$ is perpendicular to both $\nabla$ and **C**, its dot product with $\nabla$ must be zero. So if we replace **B** in Eqn (7.1) by $\nabla \times \mathbf{A}$ (where **A** is any vector at this stage) then Eqn (7.1) must still be satisfied to give

$$\mathbf{B} = \nabla \times \mathbf{A} \tag{7.3}$$

and

$$\nabla \cdot \mathbf{B} = \nabla \cdot (\nabla \times \mathbf{A}) = 0 \tag{7.4}$$

We have now found a potential function for **B**; **B** is $\nabla \times \mathbf{A}$ and so **A** is the potential function for **B** because the operation $\nabla \times$ is a differential operation. Although we have a vector potential which gives the field **B**, **A** is not yet restricted in any way. A little later we will introduce some restrictions on **A**.

## 7.2 THE SCALAR POTENTIAL $\phi$

Now consider our second Maxwell equation

$$\nabla \times \mathbf{E} = -\partial \mathbf{B}/\partial t \tag{7.5}$$

If **B** can be written as $\nabla \times \mathbf{A}$ then this equation becomes

$$\nabla \times \mathbf{E} = -\nabla \times (\partial \mathbf{A}/\partial t) \tag{7.6}$$

where we are assuming that $\mathbf{A}$ is a well behaved function of space and time so that the order of differentiation can be reversed. This equation might suggest that we could simply write $\mathbf{E} = -\partial\mathbf{A}/\partial t$ but we can be more general than this. If $\boldsymbol{\Omega}$ is a vector function such that $\nabla\times\boldsymbol{\Omega} = 0$ then we can write

$$\mathbf{E} = -\partial\mathbf{A}/\partial t+\boldsymbol{\Omega} \tag{7.7}$$

One way of making sure that $\nabla\times\boldsymbol{\Omega}$ is zero is to write $\boldsymbol{\Omega} = -\nabla\phi$ where $\phi$ is a scalar; because $\nabla\phi$ and $\nabla$ are parallel vectors their cross product must be zero. Equation (7.7) can then be written as

$$\mathbf{E} = -\partial\mathbf{A}/\partial t-\nabla\phi \tag{7.8}$$

where $\phi$ is a scalar potential function.

In the time independent (static) case, where $\partial/\partial t = 0$, Eqn (7.8) becomes

$$\mathbf{E} = -\nabla\phi$$

and so the derivative of $\phi$ gives $\mathbf{E}$.

## 7.3 THE INHOMOGENEOUS WAVE EQUATIONS IN $\phi$ AND A

The two potential functions $\mathbf{A}$ and $\phi$ are such that

$$\mathbf{B} = \nabla\times\mathbf{A}$$

and

$$\mathbf{E} = -\partial\mathbf{A}/\partial t-\nabla\phi \tag{7.10}$$

These results were obtained from the equations $\nabla\cdot\mathbf{B} = 0$ and $\nabla\times\mathbf{E} = -\partial\mathbf{B}/\partial t$ and can now be written into the two remaining Maxwell equations. The equation

$$c^2\nabla\times\mathbf{B} = \mathbf{J}/\varepsilon_0+\partial\mathbf{E}/\partial t \tag{7.11}$$

becomes

$$c^2\nabla\times(\nabla\times\mathbf{A}) = \mathbf{J}/\varepsilon_0+\frac{\partial}{\partial t}\left(-\frac{\partial\mathbf{A}}{\partial t}-\nabla\phi\right) \tag{7.12}$$

which, from the identity for the vector triple product, can be written as

$$c^2[\nabla(\nabla\cdot\mathbf{A})-\nabla^2\mathbf{A}] = \mathbf{J}/\varepsilon_0-\partial^2\mathbf{A}/\partial t^2-\nabla(\partial\phi/\partial t)$$

Rearranging terms gives

$$\nabla^2\mathbf{A}-\frac{1}{c^2}\frac{\partial^2\mathbf{A}}{\partial t^2} = -\frac{\mathbf{J}}{c^2\varepsilon_0}+\nabla\left(\frac{1}{c^2}\frac{\partial\phi}{\partial t}+\nabla\cdot\mathbf{A}\right) \tag{7.13}$$

This equation is now beginning to look like the general one dimensional wave equation which we had in Chapter 1

$$\nabla^2\psi-\frac{1}{v^2}\frac{\partial^2\psi}{\partial t^2} = 0 \tag{7.14}$$

except that we now have a term in $\mathbf{J}$ on the right hand side in addition to a term in $(1/c^2)\,\partial\phi/\partial t$ and $\nabla\cdot\mathbf{A}$. We have defined $\nabla\times\mathbf{A}$ to be $\mathbf{B}$ but as yet we have said nothing to restrict $\mathbf{A}$ and just as $\mathbf{A}$ could be any vector, $\phi$ can be any scalar at this stage. So we are free to choose these terms to be such that

$$\nabla\cdot\mathbf{A} = -\frac{1}{c^2}\frac{\partial\phi}{\partial t} \tag{7.15}$$

and under this condition Eqn (7.13) becomes

$$\boxed{\nabla^2\mathbf{A} - \frac{1}{c^2}\frac{\partial^2\mathbf{A}}{\partial t^2} = -\mathbf{J}/\varepsilon_0 c^2} \tag{7.16}$$

which is a three dimensional wave equation in $\mathbf{A}$ with a source term on the right hand side.

Each of the three vector components in Eqn (7.16) can be written as a separate scalar equation to look like Eqn (7.14) with an additional source term coming from the components of the current density $\mathbf{J}$. The condition that $\nabla\cdot\mathbf{A} = -(1/c^2)(\partial\phi/\partial t)$, taken in Eqn (7.15), is called the Lorentz gauge.

The remaining Maxwell equation is

$$\nabla\cdot\mathbf{E} = \rho/\varepsilon_0$$

which from Eqn (7.10) becomes

$$\nabla\cdot(-\partial\mathbf{A}/\partial t - \nabla\phi) = \rho/\varepsilon_0$$

or

$$-\frac{\partial}{\partial t}(\nabla\cdot\mathbf{A}) - \nabla^2\phi = \rho/\varepsilon_0$$

but we have chosen $\nabla\cdot\mathbf{A}$ to be $-(1/c^2)(\partial\phi/\partial t)$ and so we obtain

$$\boxed{\nabla^2\phi - \frac{1}{c^2}\frac{\partial^2\phi}{\partial t^2} = -\rho/\varepsilon_0} \tag{7.17}$$

which is the wave equation in $\phi$ with a source term in the charge density $\rho$ on the right hand side.

We now have two equations (Eqn (7.16) and Eqn (7.17)) which relate the $\mathbf{E}$ and $\mathbf{B}$ fields through their potential functions $\phi$ and $\mathbf{A}$ to the sources of such fields $\rho$ and $\mathbf{J}$. The next task is to find solutions for these equations which will give $\phi$ and $\mathbf{A}$ as functions of $\rho$ and $\mathbf{J}$ respectively.

## 7.4 SOLUTIONS FOR $\phi$ AND A IN TERMS OF THE FIELD SOURCES $\rho$ AND J

The inhomogeneous wave equations in $\phi$ and $\mathbf{A}$ are very similar. Each of the component scalar equations for **A** contained in Eqn (7.16) looks very like Eqn (7.17) for $\phi$, and this means that the solutions to the equations in $\phi$ and **A** must be very similar. In fact, if we find the solution for $\phi$ in terms of $\rho$ we will be able to write down the corresponding solution for **A** in terms of **J**.

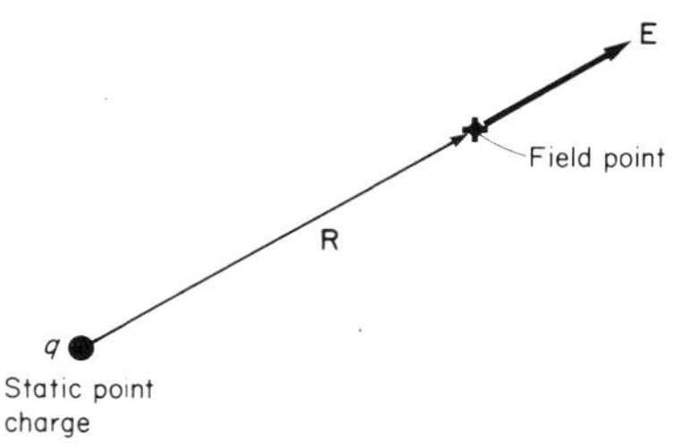


Figure 7.1

We could make a direct attempt to solve Eqn (7.17) for $\phi$, but this is a tedious process. Instead we will construct the solution starting from the equation giving the electric field **E** at a point **R** from a static point charge $q$, as shown in Fig. 7.1 (see Chapter 1, Exercise 1.1).

$$\mathbf{E}(\mathbf{R}) = \left(\frac{1}{4\pi\varepsilon_0}\right)\frac{q}{R^2}\left(\frac{\mathbf{R}}{R}\right) \tag{7.18}$$

The scalar potential $\phi$ has been defined to be such that $-\nabla\phi$ must give the field **E** in the electrostatic case. For a point charge there is spherical symmetry of the field (the field depends *only* on the distance $R$ from the charge) in which case the expression for $\nabla\phi$ in spherical polar coordinates (see Appendix B) becomes

$$\nabla\phi \equiv \frac{\partial\phi}{\partial R}\left(\frac{\mathbf{R}}{R}\right)$$

Consequently $\phi$ must be such that for a static point charge

$$\phi(\mathbf{R}) = \left(\frac{1}{4\pi\varepsilon_0}\right)\frac{q}{R} \tag{7.19}$$

We can extend this result for a point charge to a charge distribution $\rho$ in a volume $V$ by making use of the principle of superposition which says that the field due to a number of charges is given by the sum of the fields due to the individual charges.

$$\phi(\mathbf{R}) = \frac{1}{4\pi\varepsilon_0}\int_V \frac{\rho}{R}\,\mathrm{d}V \tag{7.20}$$

As we are now dealing with a spatial distribution of charge we need another vector to properly describe the spatial variation of $\rho$; say we write $\rho(\mathbf{r}_p)$ where $\mathbf{r}_p$ is now a vector from a reference point (the origin) to a point in $V$ and now write

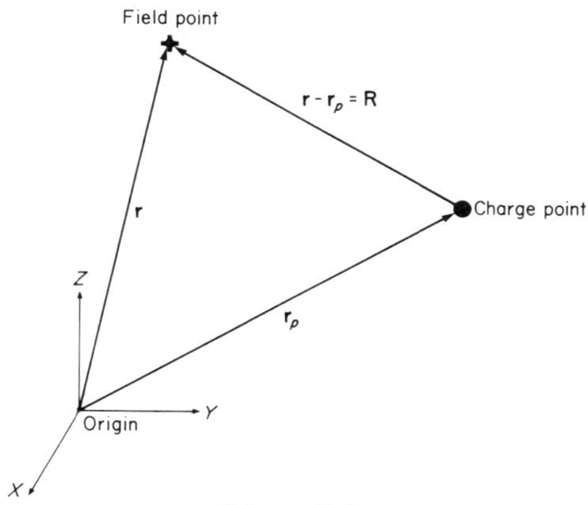


Figure 7.2

$\phi(\mathbf{r})$ where $\mathbf{r}$ is a vector from this origin to the field point as shown in Fig. 7.2. In this more general situation

$$\mathbf{R} = \mathbf{r} - \mathbf{r}_p$$

and $\phi$ is given by

$$\phi(\mathbf{r}) = \frac{1}{4\pi\varepsilon_0} \int_V \frac{\rho(\mathbf{r}_p)}{|\mathbf{r}-\mathbf{r}_p|} \, \mathrm{d}V \tag{7.21}$$

We have now constructed the solution for the *static* case and to extend this result to a moving charge distribution we must write $\phi$ and $\rho$ as functions of time and position. Since $\rho$ and $\phi$ are not at the same point and because an electromagnetic disturbance travels at a finite speed ($c$ in free space), the field at the point $\mathbf{r}$ will be due to the charge distribution at an *earlier* time (see Fig. 7.3), say

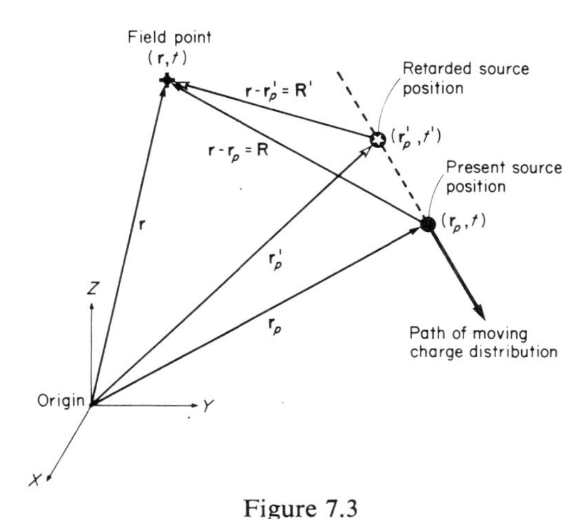


Figure 7.3

at $t'$. Therefore $\phi(\mathbf{r}, t)$ will be related to $\rho(\mathbf{r}_p', t')$ where $t'$ is the retarded time and $\mathbf{r}_p'$ is the charge position at this retarded time. The distance from the retarded source point $\mathbf{r}_p'$ to field point $\mathbf{r}$ is covered at a speed $c$ and so it takes a time $|\mathbf{r}-\mathbf{r}_p'|/c$ for the field to travel from the source point to the field point. This time delay is

$$t - t' = |\mathbf{r}-\mathbf{r}_p'|/c$$

and hence

$$t' = t - |\mathbf{r}-\mathbf{r}_p'|/c \tag{7.22}$$

Then for a moving charge distribution

$$\phi(\mathbf{r}, t) = \frac{1}{4\pi\varepsilon_0}\int_{V'} \frac{\rho(\mathbf{r}_p', t')}{|\mathbf{r}-\mathbf{r}_p'|}\,\mathrm{d}V'$$

or

$$\boxed{\phi(\mathbf{r}, t) = \frac{1}{4\pi\varepsilon_0}\int_{V'} \frac{\rho(\mathbf{r}_p', t-|\mathbf{r}-\mathbf{r}_p'|/c)}{|\mathbf{r}-\mathbf{r}_p'|}\,\mathrm{d}V'} \tag{7.23}$$

where $\mathbf{r}_p'$ is the retarded source position, $t'$ is the retarded time and the integral is evaluated over the retarded volume $V'$. This is the solution we have constructed to give the general relationship between the potential $\phi$ and the charge density $\rho$.

We can now write the corresponding expression for $\mathbf{A}$

$$\boxed{\mathbf{A}(\mathbf{r}, t) = \frac{1}{4\pi\varepsilon_0 c^2}\int_{V'} \frac{\mathbf{J}(\mathbf{r}_p', t-|\mathbf{r}-\mathbf{r}_p'|/c)}{|\mathbf{r}-\mathbf{r}_p'|}\,\mathrm{d}V'} \tag{7.24}$$

where the current density $\mathbf{J}$ must be evaluated at the retarded position $(\mathbf{r}_p', t')$ and the integral is evaluated over the retarded volume $V'$. In the expressions for $\phi$ and $\mathbf{A}$, the amount by which the time is retarded $(|\mathbf{r}-\mathbf{r}_p'|/c)$ will vary with $\mathbf{r}_p'$, that is, it depends on the point taken within the charge or current distribution. In other words, different parts of the volume $V$ are retarded by different amounts. Even in the case of a point charge this difficulty cannot be avoided when the charge is moving. At first this may seem strange, but the result for a point charge (if it is to be correct) must agree with the result which is obtained by taking the point charge as the limiting case $(V \to 0)$ for a charge distribution.

## 7.5 THE LIENARD-WIECHERT POTENTIALS

These are the scalar and vector potentials due to a moving point charge. For the proper evaluation of these potentials the point charge must be taken as a charge distribution in the limiting case where the volume approaches zero. Essentially the problem is concerned with the evaluation of the integrals for $\phi$ and $\mathbf{A}$ given in Eqns (7.23) and (7.24) above. The immediate difficulty in evaluating these integrals is to relate the retarded volume $V'$ to the present volume $V$ at time $t$. Each volume element $\mathrm{d}V'$ (at time $t'$ and position $\mathbf{r}_p'$) of the retarded volume $V'$ must be related to the corresponding volume element $\mathrm{d}V$ (at time $t$ and position $\mathbf{r}_p$) in the present volume $V$ of the moving charge or current distribution. The situation is illustrated in Fig. 7.4. We can relate the volume elements $\mathrm{d}V'$ and

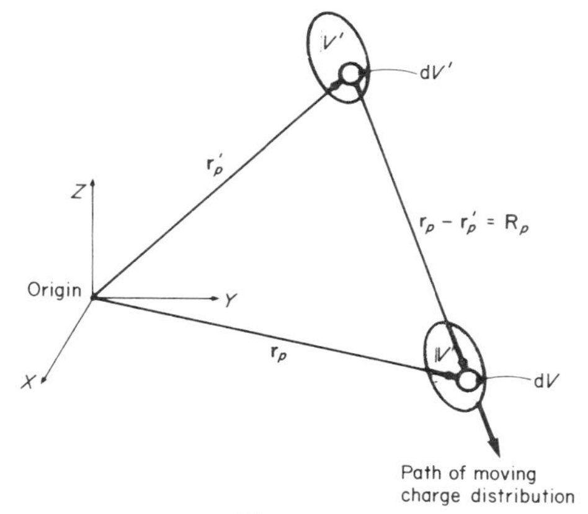


Figure 7.4

$\mathrm{d}V$ by means of the Jacobian determinant $\mathscr{J}$ (see Appendix B) which provides the transformation from one volume to the other

$$\mathrm{d}V = \mathscr{J}\,\mathrm{d}V'$$

where the Jacobian is given by

$$\mathscr{J} = \frac{\partial(x_p, y_p, z_p)}{\partial(x_p', y_p', z_p')}$$

$$= \begin{vmatrix} \dfrac{\partial x_p}{\partial x_p'} & \dfrac{\partial y_p}{\partial x_p'} & \dfrac{\partial z_p}{\partial x_p'} \\[2ex] \dfrac{\partial x_p}{\partial y_p'} & \dfrac{\partial y_p}{\partial y_p'} & \dfrac{\partial z_p}{\partial y_p'} \\[2ex] \dfrac{\partial x_p}{\partial z_p'} & \dfrac{\partial y_p}{\partial z_p'} & \dfrac{\partial z_p}{\partial z_p'} \end{vmatrix}$$

The task is now to find the derivatives in the determinant and to do this we need the general relationships between the two volumes. In the case of a moving charge distribution we assume that the distribution of charge is the same within the retarded and present volumes, that is

$$\rho(\mathbf{r}_p, t) = \rho(\mathbf{r}_p', t') \tag{7.25}$$

The relationship between the retarded and present times is given by Eqn (7.22) as

$$t' = t - \frac{|\mathbf{r}_p - \mathbf{r}_p'|}{c}$$

We can say that the vector, $\mathbf{R}_p$, joining corresponding elements from the retarded to the present volume, is given by

$$\mathbf{R}_p = \mathbf{r}_p - \mathbf{r}_p' \tag{7.26}$$

and this vector must be related to the time difference $t-t'$. For the $x$-component of $\mathbf{r}_p - \mathbf{r}_p'$ we can write

$$x_p - x_p' = f(t-t') \tag{7.27}$$

and expand the function of $(t-t')$ as a Taylor series

$$f(t-t') = a_0 + a_1(t-t') + a_2(t-t')^2 + \cdots$$

The coefficients of the power series in $(t-t')$ are related to the value of the function and its derivatives at $t = t'$,

$$a_0 = f \quad (\text{at} \quad t = t') \quad \text{where} \quad f \equiv f(t-t')$$

$$a_1 = f' \quad (\text{at} \quad t = t') \quad \text{where} \quad f' \equiv \frac{\partial}{\partial t}\{f(t-t')\}$$

$$a_2 = \frac{f''}{2!} \quad (\text{at} \quad t = t') \quad \text{where} \quad f'' \equiv \frac{\partial^2}{\partial t^2}\{f(t-t')\}$$

and in general

$$a_n = \frac{f^n}{n!} \quad (\text{at} \quad t = t') \quad \text{where} \quad f^n \equiv \frac{\partial^n}{\partial t^n}\{f(t-t')\}$$

Then

$$x_p - x_p' = [f\,(\text{at } t = t')] + [f'\,(\text{at } t = t')](t-t') + [f''\,(\text{at } t = t')](t-t')^2/2 + \cdots + [f^n\,(\text{at } t = t')](t-t')^n/n! + \cdots \tag{7.28}$$

The value of the function $[x_p(t) - x_p'(t')]$ at $t = t'$ is zero because at $t = t'$, $x_p = x_p'$ and so the first term in the series is zero. The coefficient of the second term is the function differentiated with respect to time $t$ and evaluated at $t = t'$, but

$$\frac{\partial}{\partial t}[x_p(t) - x_p'(t')] = \frac{\partial x_p}{\partial t}$$

and it is clear that the derivative is the $x$-component of the velocity $\mathbf{v}$ of the charge distribution, that is

$$v_x(t') = f'\,(\text{at } t = t')$$

The second derivative of $[x_p(t) - x_p'(t')]$ gives the derivative of the $x$-component of the velocity and therefore

$$\dot{v}_x(t') = f'' \text{ (at } t = t'\text{)}$$

Then the power series can be written as

$$x_p - x_p' = v_x(t')(t-t') + \dot{v}_x(t')(t-t')^2/2 + \ddot{v}_x(t')(t-t')^3/3! + \cdots \quad (7.29)$$

and this expression can be used to obtain some of the terms in the Jacobian. In a similar way we can write

$$y_p - y_p' = v_y(t')(t-t') + \dot{v}_y(t')(t-t')^2/2 + \ddot{v}_y(t')(t-t')^3/3! + \cdots \quad (7.30)$$

and

$$z_p - z_p' = v_z(t')(t-t') + \dot{v}_z(t')(t-t')^2/2 + \ddot{v}_z(t')(t-t')^3/3! + \cdots \quad (7.31)$$

Differentiating Eqn (7.29) with respect to $x_p'$ and remembering that the velocity terms are functions of the retarded position coordinates gives

$$\begin{aligned}\frac{\partial x_p}{\partial x_p'} - 1 = &-v_x(t')\left(\frac{\partial t'}{\partial x_p'}\right) + \dot{v}_x(t')\left(\frac{\partial t'}{\partial x_p'}\right)(t-t') \\ &- \dot{v}_x(t')(t-t')\left(\frac{\partial t'}{\partial x_p'}\right) + \frac{\ddot{v}_x(t')(t-t')^2}{2!}\left(\frac{\partial t'}{\partial x_p'}\right) \\ &- \frac{\ddot{v}_x(t')(t-t')^2}{2!}\left(\frac{\partial t'}{\partial x_p'}\right) + \frac{\dddot{v}_x(t')(t-t')^3}{3!}\left(\frac{\partial t'}{\partial x_p'}\right) \\ &+ \cdots \end{aligned} \quad (7.32)$$

With the exception of the first term, all the terms on the right hand side cancel to give

$$\frac{\partial x_p}{\partial x_p'} - 1 = -v_x(t')\left(\frac{\partial t'}{\partial x_p'}\right) \quad (7.33)$$

In a similar way, differentiation of Eqn (7.29) with respect to $y_p'$ gives

$$\frac{\partial x_p}{\partial y_p'} = -v_x(t')\left(\frac{\partial t'}{\partial y_p'}\right) \quad (7.34)$$

and differentiation of Eqn (7.29) with respect to $z_p'$ gives

$$\frac{\partial x_p}{\partial z_p'} = -v_x(t')\left(\frac{\partial t'}{\partial z_p'}\right) \quad (7.35)$$

Similar results are obtained from Eqns (7.30) and (7.31) for the derivatives of $y_p$ and $z_p$ with respect to the retarded space coordinates. All of these expressions contain (as do Eqns (7.33), (7.34) and (7.35)) the derivatives of the retarded time with respect to the retarded coordinates, and these terms can be found from the relationship

$$t' = t - \frac{|\mathbf{r}_p - \mathbf{r}_p'|}{c}$$

where

$$\mathbf{r}_p - \mathbf{r}_p' = (x_p - x_p')\mathbf{i} + (y_p - y_p')\mathbf{j} + (z_p - z_p')\mathbf{k}$$

and

$$|\mathbf{r}_p - \mathbf{r}_p'| = \sqrt{(x_p - x_p')^2 + (y_p - y_p')^2 + (z_p - z_p')^2} = R_p$$

hence

$$\frac{\partial |\mathbf{r}_p - \mathbf{r}_p'|}{\partial x_p'} = \left(\frac{1}{2}\right)\{(x_p - x_p')^2 + (y_p - y_p')^2 + (z_p - z_p')^2\}^{-\frac{1}{2}}(2)(x_p - x_p')(-1)$$

$$= -\frac{(x_p - x_p')}{R_p}$$

$$= -n_{px}'$$

where $n_{px}'$ is the magnitude of the $x$-component of the unit vector in the direction of $\mathbf{r}_p - \mathbf{r}_p'$. Therefore

$$\frac{\partial t'}{\partial x_p'} = \frac{n_{px}'}{c}$$

and similarly

$$\frac{\partial t'}{\partial y_p'} = \frac{n_{py}'}{c}$$

$$\frac{\partial t'}{\partial z_p'} = \frac{n_{pz}'}{c}$$

We can now obtain expressions for each of the terms in the Jacobian

$$\frac{\partial x_p}{\partial x_p'} = 1 - \frac{v_x(t')n_{px}'}{c}, \quad \frac{\partial x_p}{\partial y_p'} = -\frac{v_x(t')n_{py}'}{c}$$

$$\frac{\partial x_p}{\partial z_p'} = -\frac{v_x(t')n_{pz}'}{c}$$

$$\frac{\partial y_p}{\partial y_p'} = 1 - \frac{v_y(t')n_{py}'}{c}, \quad \frac{\partial y_p}{\partial x_p'} = -\frac{v_y(t')n_{px}'}{c}$$

$$\frac{\partial y_p}{\partial z_p'} = -\frac{v_y(t')n_{pz}'}{c}$$

$$\frac{\partial z_p}{\partial z_p'} = 1 - \frac{v_z(t')n_{pz}'}{c}, \quad \frac{\partial z_p}{\partial y_p'} = -\frac{v_z(t')n_{py}'}{c}$$

$$\frac{\partial z_p}{\partial x_p'} = -\frac{v_z(t')n_{px}'}{c}$$

On substituting our expressions for the terms in the Jacobian and expanding the determinant (this is left as an exercise, and the full working is given in Appendix 7.1 at the end of this chapter) we obtain

$$\mathscr{J} = 1 - \frac{v_x(t')n'_{px}}{c} - \frac{v_y(t')n'_{py}}{c} - \frac{v_z(t')n'_{pz}}{c}$$

that is

$$\boxed{\mathscr{J} = 1 - \mathbf{v}(t')\cdot\mathbf{n}'_p/c} \tag{7.36}$$

Then the relationship between the volume element $\mathrm{d}V$ and the retarded volume element $\mathrm{d}V'$ is

$$\mathrm{d}V = [1 - \mathbf{v}(t')\cdot\mathbf{n}'_p/c]\,\mathrm{d}V'$$

The expression for the scalar potential $\phi(\mathbf{r}, t)$

$$\phi(\mathbf{r}, t) = \frac{1}{4\pi\varepsilon_0}\int_{V'} \frac{\rho(\mathbf{r}'_p, t')}{|\mathbf{r} - \mathbf{r}'_p|}\,\mathrm{d}V'$$

can now be transformed from an integral over the retarded volume $V'$ into the corresponding integral over the volume $V$. To obtain the result for a *point charge* we let the volume approach zero in which case $|\mathbf{r} - \mathbf{r}'_p|$ will be constant over the integration, $R'$ (say). We also have the condition given in Eqn (7.25) that

$$\rho(\mathbf{r}_p, t) = \rho(\mathbf{r}'_p, t')$$

and we obtain

$$\phi(\mathbf{r}, t) = \frac{1}{4\pi\varepsilon_0}\int_{\substack{V \\ V\to 0}} \frac{\rho(\mathbf{r}_p, t)}{R'}\left\{\frac{\mathrm{d}V}{[1 - \mathbf{v}(t')\cdot\mathbf{n}'_p/c]}\right\}$$

that is

$$\boxed{\phi(\mathbf{r}, t) = \left(\frac{1}{4\pi\varepsilon_0}\right)\frac{q}{R'[1 - \mathbf{v}(t')\cdot\mathbf{n}'_p/c]}} \tag{7.37}$$

for a point charge $q$, where $R' = |\mathbf{r} - \mathbf{r}'_p|$ and all primed quantities are to be evaluated at the retarded charge position $(\mathbf{r}'_p, t')$. In a similar way we obtain the vector potential for a moving point charge as

$$\boxed{\mathbf{A}(\mathbf{r}, t) = \left(\frac{1}{4\pi\varepsilon_0 c^2}\right)\frac{q\mathbf{v}(t')}{R'[1 - \mathbf{v}(t')\cdot\mathbf{n}'_p/c]}} \tag{7.38}$$

Equations (7.37) and (7.38) are the Lienard-Wiechert potentials for a moving point charge.

# APPENDIX 7.1

**To Expand the Jacobian Determinant and Show that $\mathscr{J} = 1 - \mathbf{v}(t') \cdot \mathbf{n}'_p / c$**

*Solution*

$$\mathscr{J} = \begin{vmatrix} \dfrac{\partial x_p}{\partial x'_p} & \dfrac{\partial y_p}{\partial x'_p} & \dfrac{\partial z_p}{\partial x'_p} \\[2ex] \dfrac{\partial x_p}{\partial y'_p} & \dfrac{\partial y_p}{\partial y'_p} & \dfrac{\partial z_p}{\partial y'_p} \\[2ex] \dfrac{\partial x_p}{\partial z'_p} & \dfrac{\partial y_p}{\partial z'_p} & \dfrac{\partial z_p}{\partial z'_p} \end{vmatrix}$$

$$\begin{aligned} = \quad & \frac{\partial x_p}{\partial x'_p}\left(\frac{\partial y_p}{\partial y'_p}\frac{\partial z_p}{\partial z'_p} - \frac{\partial z_p}{\partial y'_p}\frac{\partial y_p}{\partial z'_p}\right) \\ & - \frac{\partial y_p}{\partial x'_p}\left(\frac{\partial x_p}{\partial y'_p}\frac{\partial z_p}{\partial z'_p} - \frac{\partial z_p}{\partial y'_p}\frac{\partial x_p}{\partial z'_p}\right) \\ & + \frac{\partial z_p}{\partial x'_p}\left(\frac{\partial x_p}{\partial y'_p}\frac{\partial y_p}{\partial z'_p} - \frac{\partial y_p}{\partial y'_p}\frac{\partial x_p}{\partial z'_p}\right) \end{aligned}$$

$$\begin{aligned} = & \left(1 - \frac{v'_x n'_{px}}{c}\right)\left[\left(1 - \frac{v'_y n'_{py}}{c}\right)\left(1 - \frac{v'_z n'_{pz}}{c}\right) - \left(-\frac{v'_z n'_{py}}{c}\right)\left(-\frac{v'_y n'_{pz}}{c}\right)\right] \\ & - \left(-\frac{v'_y n'_{px}}{c}\right)\left[\left(-\frac{v'_x n'_{py}}{c}\right)\left(1 - \frac{v'_z n'_{pz}}{c}\right) - \left(-\frac{v'_z n'_{py}}{c}\right)\left(-\frac{v'_x n'_{pz}}{c}\right)\right] \\ & + \left(-\frac{v'_z n'_{px}}{c}\right)\left[\left(-\frac{v'_x n'_{py}}{c}\right)\left(-\frac{v'_y n'_{pz}}{c}\right) - \left(1 - \frac{v'_y n'_{py}}{c}\right)\left(-\frac{v'_x n'_{pz}}{c}\right)\right] \end{aligned}$$

where $v' \equiv v(t')$

$$\therefore \qquad \mathscr{J} = \left(1 - \frac{v'_x n'_{px}}{c}\right)\left[1 - \frac{v'_z n'_{pz}}{c} - \frac{v'_y n'_{py}}{c} + \frac{v'_y v'_z n'_{py} n'_{pz}}{c^2} - \frac{v'_y v'_z n'_{py} n'_{pz}}{c^2}\right]$$

$$- \left(-\frac{v'_y n'_{px}}{c}\right)\left[-\frac{v'_x n'_{py}}{c} + \frac{v'_x v'_z n'_{py} n'_{pz}}{c^2} - \frac{v'_x v'_y n'_{py} n'_{pz}}{c^2}\right]$$

$$+ \left(-\frac{v'_z n'_{px}}{c}\right)\left[\frac{v'_x v'_y n'_{py} n'_{pz}}{c^2} + \frac{v'_x n'_{pz}}{c} - \frac{v'_x v'_y n'_{py} n'_{pz}}{c^2}\right]$$

$$= 1 - \frac{v'_z n'_{pz}}{c} - \frac{v'_y n'_{py}}{c} - \frac{v'_x n'_{px}}{c} + \frac{v'_x v'_z n'_{px} n'_{pz}}{c^2}$$

$$+ \frac{v'_x v'_y n'_{px} n'_{py}}{c^2} - \frac{v'_x v'_y n'_{px} n'_{py}}{c^2} - \frac{v'_x v'_z n'_{px} n'_{pz}}{c^2}$$

that is

$$\boxed{\mathscr{J} = 1 - \mathbf{v}(t') \cdot \mathbf{n}'_p / c}$$

# Chapter 8

# SIMPLE RADIATING SYSTEMS

In Chapter 7 we found the scalar and vector potentials due to a moving point charge (the Lienard-Wiechert potentials) and we now use these potentials to find the fields produced by the movement of charge in certain simple situations. We start by considering the radiation due to a free charge being driven by a field which has a sinusoidal variation in time. These results for a free charge are readily extended to the case of a bound charge and they also form the basis for our consideration of the radiation from simple antenna systems.

## 8.1 RADIATION FROM A SLOWLY MOVING ACCELERATED POINT CHARGE

We consider a point charge $q$ in free space where it is free to move under the influence of a linearly polarized plane electromagnetic wave described by

$$\mathbf{E} = \mathbf{E}_0 \exp[i(\omega t - k_x x)] \tag{8.1}$$

where $\mathbf{E}_0$ is a constant vector which is perpendicular to the $x$-axis and where $\omega$ and $k_x$ are constants. The force on a free charge in an electromagnetic field is given by

$$\mathbf{F} = q(\mathbf{E} + \mathbf{v} \times \mathbf{B}) \tag{8.2}$$

If the velocity, $\mathbf{v}$, of the charge is such that $v \ll c$ then the magnetic force is much less than the electric force and the term $q\mathbf{v} \times \mathbf{B}$ can be neglected. Choosing the constant vector $\mathbf{E}_0$ to be in the $z$-direction means that the acceleration of the charge will also be in this direction (see Fig. 8.1) and the motion of the charge is described by

$$m\ddot{z} = qE_0 \cos \omega t \tag{8.3}$$

where $m$ is the mass of the charge. The steady state solution to this differential equation is

$$z = -qE_0(\cos \omega t)/m\omega^2 \tag{8.4}$$

and the velocity of the charge is then

$$\dot{z} = qE_0(\sin \omega t)/m\omega \tag{8.5}$$

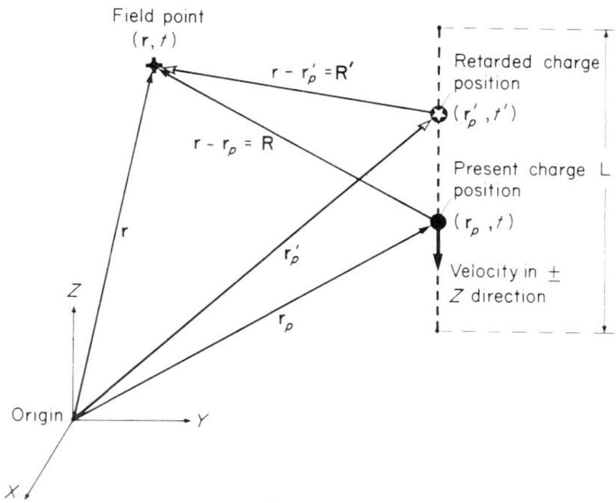


Figure 8.1

The retarded velocity $\mathbf{v}(t')$ is obtained by making the substitution

$$t' = t - R'/c \tag{8.6}$$

where $R'$ is the distance from the field point $(\mathbf{r}, t)$ to the retarded charge position $(\mathbf{r}'_p, t')$, that is

$$R' = |\mathbf{r} - \mathbf{r}'_p| \tag{8.7}$$

which gives

$$\mathbf{v}(t') = \mathbf{k}\,\frac{qE_0}{m\omega}\sin[\omega(t - R'/c)] \tag{8.8}$$

The Lienard-Wiechert potential is

$$\mathbf{A}(\mathbf{r}, t) = \mathbf{k}\,\frac{q^2 E_0 \sin[\omega(t - R'/c)]}{m\omega 4\pi\varepsilon_0 c^2 R'[1 - \mathbf{v}(t')\cdot\mathbf{n}'/c]} \tag{8.9}$$

but to neglect the force due to the magnetic field we have already taken $v \ll c$ and therefore we can write

$$\mathbf{A}(\mathbf{r}, t) = \mathbf{k}\,\frac{q^2 E_0 \sin[\omega(t - R'/c)]}{m\omega 4\pi\varepsilon_0 c^2 R'} \tag{8.10}$$

If we now impose the further restriction that we confine our attention to the fields at a *large* distance from the charge this means that, in comparison to this large distance, the separation of the charge position from its retarded position is negligible

$$\begin{aligned} R' &= |\mathbf{r} - \mathbf{r}'_p| \\ &\simeq |\mathbf{r} - \mathbf{r}_p| \\ &= R \end{aligned} \tag{8.11}$$

This new condition needs to be examined a little more closely. We are saying that the distance $R$ from the field point $(\mathbf{r}, t)$ to the charge point $(\mathbf{r}_p, t)$ is much greater than the distance between the retarded and present positions of the charge. Since the motion of the charge is sinusoidal, the maximum difference $L$ between any two positions is twice the amplitude. The amplitude is given by Eqn (8.4) from which we can write $L = 2qE_0/m\omega^2$ and this, from Eqn (8.5), is $L = 2v_0/\omega$ where $v_0$ is the maximum velocity. The condition is then

$$R \gg 2v_0/\omega$$

but we have already said that $v \ll c$ and therefore our condition is fulfilled if

$$R \gg 2c/\omega$$

or, since $c = \lambda\omega/2\pi$ (where $\lambda$ is the wavelength)

$$\boxed{\begin{aligned} R &\gg \lambda \\ L &\ll \lambda \end{aligned}} \tag{8.12}$$

This then is the condition which expresses the restriction to large distances from the charge and under this condition we can write Eqn (8.10) as

$$\mathbf{A}(\mathbf{r}, t) = \mathbf{k}\left(\frac{q^2 E_0}{m\omega}\right) \frac{\sin[\omega(t - R/c)]}{4\pi\varepsilon_0 c^2 R} \tag{8.13}$$

where

$$\begin{aligned} \mathbf{R} &= \mathbf{r} - \mathbf{r}_p \\ &= (x - x_p)\mathbf{i} + (y - y_p)\mathbf{j} + (z - z_p)\mathbf{k} \end{aligned} \tag{8.14}$$

and

$$R = \sqrt{(x - x_p)^2 + (y - y_p)^2 + (z - z_p)^2} \tag{8.15}$$

The magnetic field $\mathbf{B}$ due to the motion of the charge is given by

$$\mathbf{B} = \nabla \times \mathbf{A} = \begin{vmatrix} \mathbf{i} & \mathbf{j} & \mathbf{k} \\ \dfrac{\partial}{\partial x} & \dfrac{\partial}{\partial y} & \dfrac{\partial}{\partial z} \\ 0 & 0 & A \end{vmatrix}$$

that is

$$\mathbf{B} = \mathbf{i}\frac{\partial A}{\partial y} - \mathbf{j}\frac{\partial A}{\partial x} + \mathbf{k}(0) \tag{8.16}$$

To obtain the $y$ and $x$ derivatives of $A$, we find $\partial A/\partial R$ from Eqn (8.13) and find

$\partial R/\partial y$ and $\partial R/\partial x$ from Eqn (8.15); multiplication then gives $\partial A/\partial y$ and $\partial A/\partial x$.

$$\frac{\partial A}{\partial R} = \frac{q^2 E_0}{4\pi\varepsilon_0 c^2 m\omega}\left\{\frac{R\left(-\frac{\omega}{c}\right)\cos[\omega(t-R/c)] - \sin[\omega(t-R/c)]}{R^2}\right\}$$

Since we are restricted to large values of $R$, the difference in the relative magnitudes of the coefficients means that the sine term is negligible in comparison to the cosine term ($R \gg \lambda$ is equivalent to $R\omega/c \gg 1$), that is

$$\frac{\partial A}{\partial R} = \frac{q^2 E_0}{4\pi\varepsilon_0 c^2 m\omega}\left\{\frac{\left(-\frac{\omega}{c}\right)\cos[\omega(t-R/c)]}{R}\right\} \quad \text{for} \quad R \gg \lambda \qquad (8.17)$$

From Eqn (8.15)

$$\frac{\partial R}{\partial x} = \frac{(x-x_p)}{R} \qquad (8.18)$$

and

$$\frac{\partial R}{\partial y} = \frac{(y-y_p)}{R} \qquad (8.19)$$

Therefore

$$\frac{\partial A}{\partial y} = \left(\frac{\partial A}{\partial R}\right)\left(\frac{\partial R}{\partial y}\right) = \left(-\frac{q^2 E_0}{4\pi\varepsilon_0 c^3 m}\right)\frac{\cos[\omega(t-R/c)]}{R^2}(y-y_p)$$

$$\frac{\partial A}{\partial x} = \left(\frac{\partial A}{\partial R}\right)\left(\frac{\partial R}{\partial x}\right) = \left(-\frac{q^2 E_0}{4\pi\varepsilon_0 c^3 m}\right)\frac{\cos[\omega(t-R/c)]}{R^2}(x-x_p)$$

and Eqn (8.16) for **B** becomes

$$\mathbf{B} = \left(-\frac{q^2 E_0}{4\pi\varepsilon_0 c^3 m}\right)\frac{\cos[\omega(t-R/c)]}{R^2}\{(y-y_p)\mathbf{i} - (x-x_p)\mathbf{j} + \mathbf{k}(0)\} \qquad (8.20)$$

but

$$\mathbf{k}\times\mathbf{R} = \begin{vmatrix} \mathbf{i} & \mathbf{j} & \mathbf{k} \\ 0 & 0 & 1 \\ (x-x_p) & (y-y_p) & (z-z_p) \end{vmatrix}$$

$$= -\{\mathbf{i}(y-y_p) - \mathbf{j}(x-x_p) + \mathbf{k}(0)\} \qquad (8.21)$$

and therefore we can write **B** as

$$\boxed{\mathbf{B} = \left(\frac{q^2 E_0}{4\pi\varepsilon_0 c^3 m}\right)\frac{\cos[\omega(t-R/c)]}{R}\left(\mathbf{k}\times\frac{\mathbf{R}}{R}\right)} \qquad (8.22)$$

This result shows that at a large distance from the charge the magnetic field is perpendicular to the movement of the charge and perpendicular to the line from the charge to the field point. The magnitude of the vector cross product increases as the angle between **k** and **R** approaches 90° and so (for a given value of $R$) has a maximum value in the plane containing the charge and at right angles to the direction in which the charge is moving; the vector cross product term shows that there is zero field along the line of motion of the charge.

Having found the **B** field we can use the Maxwell relation

$$c^2\nabla\times\mathbf{B} = \partial\mathbf{E}/\partial t \quad \text{(for free space)} \tag{8.23}$$

to find the **E** field

$$\nabla\times\mathbf{B} = \begin{vmatrix} \mathbf{i} & \mathbf{j} & \mathbf{k} \\ \dfrac{\partial}{\partial x} & \dfrac{\partial}{\partial y} & \dfrac{\partial}{\partial z} \\ B_x & B_y & 0 \end{vmatrix}$$

where in this case we have seen that the $x$-component of the magnetic field is

$$B_x = \left(\frac{-q^2E_0}{4\pi\varepsilon_0 c^3 m}\right)\frac{\cos[\omega(t-R/c)]}{R^2}(y-y_p)$$

and the $y$-component is

$$B_y = -\left(\frac{-q^2E_0}{4\pi\varepsilon_0 c^3 m}\right)\frac{\cos[\omega(t-R/c)]}{R^2}(x-x_p)$$

Then

$$\begin{aligned}
\frac{\partial\mathbf{E}}{\partial t} &= c^2\left\{\mathbf{i}\left(-\frac{\partial B_y}{\partial z}\right)-\mathbf{j}\left(-\frac{\partial B_x}{\partial z}\right)+\mathbf{k}\left(\frac{\partial B_y}{\partial x}-\frac{\partial B_x}{\partial y}\right)\right\} \\
&= \left(\frac{-q^2E_0}{4\pi\varepsilon_0 cm}\right)\left\{\mathbf{i}(x-x_p)\left(\frac{\partial R}{\partial z}\right)\frac{\partial}{\partial R}\left(\frac{\cos[\omega(t-R/c)]}{R^2}\right)\right. \\
&\quad +\mathbf{j}(y-y_p)\left(\frac{\partial R}{\partial z}\right)\frac{\partial}{\partial R}\left(\frac{\cos[\omega(t-R/c)]}{R^2}\right) \\
&\quad -\mathbf{k}\left[\frac{\cos[\omega(t-R/c)]}{R^2}+(x-x_p)\left(\frac{\partial R}{\partial x}\right)\frac{\partial}{\partial R}\left(\frac{\cos[\omega(t-R/c)]}{R^2}\right)\right. \\
&\quad \left.\left.+\frac{\cos[\omega(t-R/c)]}{R^2}+(y-y_p)\left(\frac{\partial R}{\partial y}\right)\frac{\partial}{\partial R}\left(\frac{\cos[\omega(t-R/c)]}{R^2}\right)\right]\right\}
\end{aligned}$$

Completing the differentiation and applying the condition $(R\omega/c) \gg 1$ gives

$$\frac{\partial \mathbf{E}}{\partial t} = \left(\frac{-q^2 E_0}{4\pi\varepsilon_0 cm}\right)\left\{\mathbf{i}(x-x_p)(z-z_p)\left(-\frac{\omega}{c}\right)\left[\frac{-\sin[\omega(t-R/c)]}{R^3}\right]\right.$$

$$+\mathbf{j}(y-y_p)(z-z_p)\left(-\frac{\omega}{c}\right)\left[\frac{-\sin[\omega(t-R/c)]}{R^3}\right]$$

$$\left.-\mathbf{k}[(x-x_p)^2+(y-y_p)^2]\left(-\frac{\omega}{c}\right)\left[\frac{-\sin[\omega(t-R/c)]}{R^3}\right]\right\}$$

Therefore

$$\mathbf{E} = \left(\frac{q^2 E_0}{4\pi\varepsilon_0 cm}\right)\left(\frac{1}{c}\right)\frac{\cos[\omega(t-R/c)]}{\mathrm{R}^3}\{\mathbf{i}(x-x_p)(z-z_p)$$

$$+\mathbf{j}(y-y_p)(z-z_p)-\mathbf{k}[(y-y_p)^2+(x-x_p)^2]\}$$

From Eqn (8.21) we can write

$$(\mathbf{k}\times\mathbf{R})\times\mathbf{R} = \begin{vmatrix} \mathbf{i} & \mathbf{j} & \mathbf{k} \\ -(y-y_p) & (x-x_p) & 0 \\ (x-x_p) & (y-y_p) & (z-z_p) \end{vmatrix}$$

$$= \{\mathbf{i}(x-x_p)(z-z_p)+\mathbf{j}(y-y_p)(z-z_p)-\mathbf{k}[(y-y_p)^2+(x-x_p)^2]\}$$

Therefore

$$\boxed{\mathbf{E} = \left(\frac{q^2 E_0}{4\pi\varepsilon_0 c^2 m}\right)\frac{\cos[\omega(t-R/c)]}{R}\left\{\left(\mathbf{k}\times\frac{\mathbf{R}}{R}\right)\times\frac{\mathbf{R}}{R}\right\}} \qquad (8.24)$$

Equations (8.22) and (8.24) show that the **E** and **B** fields are mutually perpendicular and that they are perpendicular to the direction $\mathbf{R}/R$. The electromagnetic field due to the sinusoidal oscillation of the charge is seen to be a transverse electromagnetic wave propagating radially outward from the moving charge. The **E** and **B** fields have their maximum values in the plane containing the charge and perpendicular to the motion of the charge. Our results apply only to the case where the velocity of the charge is much less than $c$ and only at large $(R \gg \lambda)$ distances from the charge.

## 8.2 ENERGY SCATTERED BY A FREE CHARGE

We have found the **E** and **B** fields in the wave radiated from a free charge which is being driven by the fields in a plane electromagnetic wave incident on the charge. In Chapter 1 it was shown that the Poynting vector $\mathbf{S} = \varepsilon_0 c^2 \mathbf{E}\times\mathbf{B}$ represents the rate at which energy is flowing per unit area in the direction of **S** at a point in the field.

Equations (8.22) and (8.24) show that the magnitude of both **B** and **E** varies with sin $\theta$ where $\theta$ is the angle between **R** and the direction **k** of the charge velocity. The vector $\mathbf{E}\times\mathbf{B}$ is perpendicular to both **E** and **B** and consequently **S** points radially outward from the charge, in the direction $\mathbf{R}/R$.

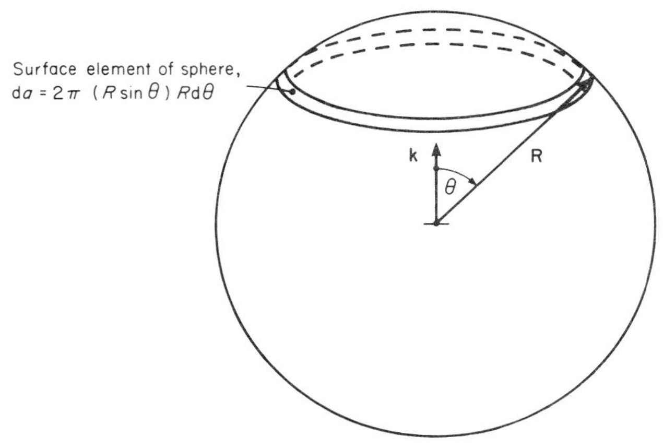


Figure 8.2

We measure the energy radiated by the charge in terms of the average rate at which energy crosses a spherical surface of radius $R$ drawn (see Fig. 8.2) so that the charge moves about the centre of the spherical surface. If the radius of the sphere is taken to be much greater than the amplitude of the motion of the charge, $R$ can then be taken as the distance of the charge from the spherical surface. At a point on the surface of the sphere the Poynting vector is

$$\mathbf{S} = \varepsilon_0 c^2 \mathbf{E}\times\mathbf{B}$$

where from Eqns (8.22) and (8.24) we can write

$$B = \left(\frac{q^2 E_0}{4\pi\varepsilon_0 c^3 m}\right)\frac{\cos[\omega(t-R/c)]}{R}\sin\theta$$

and

$$E = \left(\frac{q^2 E_0}{4\pi\varepsilon_0 c^2 m}\right)\frac{\cos[\omega(t-R/c)]}{R}\sin\theta$$

therefore

$$\mathbf{S} = \varepsilon_0 c^2\left(\frac{q^4 E_0^2}{16\pi^2\varepsilon_0^2 c^5 m^2}\right)\frac{\cos^2[\omega(t-R/c)]}{R^2}\sin^2\theta\left(\frac{\mathbf{R}}{R}\right)$$

The time average value of **S** is obtained by finding the time average of $\cos^2[\omega(t-R/c)]$, which is shown in Appendix B to be $\frac{1}{2}$. Therefore the time average Poynting vector $\overline{\mathbf{S}}$ is,

$$\overline{\mathbf{S}} = \left(\frac{q^4 E_0^2}{32\pi^2\varepsilon_0 c^3 m^2}\right)\left(\frac{\sin^2\theta}{R^2}\right)\left(\frac{\mathbf{R}}{R}\right)$$

The time average rate at which energy crosses a surface element d$a$ is $\bar{\mathbf{S}}\cdot\mathbf{n}\,\mathrm{d}a$ where $\mathbf{n}$ is the unit normal to d$a$ and consequently the total time average rate at which energy flows outward across the sphere of radius $R$ is

$$W_0 = \int_0^\pi \left(\frac{q^4E_0^2}{32\pi^2\varepsilon_0c^3m^2}\right)\left(\frac{\sin^2\theta}{R^2}\right)2\pi(R\sin\theta)R\,\mathrm{d}\theta$$

$$= \frac{q^4E_0^2}{16\pi\varepsilon_0c^3m^2}\int_0^\pi \sin^3\theta\,\mathrm{d}\theta$$

but

$$\int_0^\pi \sin^3\theta\,\mathrm{d}\theta = 4/3$$

and therefore

$$W_0 = \frac{q^4E_0^2}{12\pi\varepsilon_0c^3m^2}\ \mathrm{J\ s^{-1}} \tag{8.25}$$

The Poynting vector for the wave driving the charge is also given by

$$\mathbf{S} = \varepsilon_0c^2\mathbf{E}\times\mathbf{B}$$

and since this is a transverse electromagnetic wave in free space, the time average is

$$\bar{S} = \varepsilon_0c^2(\tfrac{1}{2})(E_0)\left(\frac{1}{c}E_0\right)$$

or

$$W_i = \frac{\varepsilon_0c}{2}E_0^2\ \mathrm{J\ s^{-1}\ m^{-2}} \tag{8.26}$$

where $W_i$ is the time average rate of flow of energy per unit area in the incoming wave which is driving the charge. The total scattering cross-section $\sigma_s$ for a charge $q$ is defined as the ratio of the total average rate of flow of scattered radiation ($W_0$) to the average rate of flow of energy per unit area in the incoming wave ($W_i$).

$$\sigma_s = \frac{W_0}{W_i}$$

$$\sigma_s = \left(\frac{q^4E_0^2}{12\pi\varepsilon_0c^3m^2}\right)\Big/\left(\frac{\varepsilon_0cE_0^2}{2}\right)$$

that is

$$\boxed{\sigma_s = \frac{q^4}{6\pi\varepsilon_0^2c^4m^2}\ \mathrm{m^2}} \tag{8.27}$$

This is the Thomson formula for the scattering cross-section of a free charge $q$.

## 8.3 SCATTERING OF RADIATION BY A BOUND CHARGE

In the simple 'molecular' model given in Chapter 3 the bound molecular charge is allowed to move with damped oscillations governed by a force equation of the type

$$qE = m(\ddot{z}+\gamma\dot{z}+\omega_0^2 z)$$

where the time variation of the driving force is

$$E = E_0 \cos \omega t$$

The steady state solution to the equation

$$\left(\frac{qE_0}{m}\right) \cos \omega t = \ddot{z}+\gamma\dot{z}+\omega_0^2 z$$

is

$$z = \left\{\left(\frac{qE_0}{m}\right) \Big/ \sqrt{\omega^2\gamma^2+(\omega^2-\omega_0^2)^2}\right\} \cos(\omega t-\beta)$$

where

$$\beta = \tan^{-1}\left(\frac{\omega\gamma}{\omega_0^2-\omega^2}\right)$$

To calculate the scattering cross-section for such a bound charge we are concerned only with time averaging the energy flows and the phase angle $\beta$ is of no importance; then for this purpose we can write the displacement of the bound charge as

$$z = \left\{\left(\frac{qE_0}{m}\right) \Big/ \sqrt{\omega^2\gamma^2+(\omega^2-\omega_0^2)^2}\right\} \cos \omega t \tag{8.28}$$

The corresponding result for a free charge was given in Eqn (8.4) as

$$z = -\left\{\left(\frac{qE_0}{m}\right) \Big/ \omega^2\right\} \cos \omega t$$

where the negative sign is not of importance as we are not concerned with relative phases. The similarity between these expressions for the displacement means that we can convert the results for the free charge into the corresponding results for the bound charge by simply multiplying each term in $E_0$ by

$$\omega^2/\sqrt{\omega^2\gamma^2+(\omega^2-\omega_0^2)^2}$$

Then from Eqn (8.25) we can write down the result for the time averaged total energy radiated per second by a bound charge as

$$W_0' = \left(\frac{q^4E_0^2}{12\pi\varepsilon_0 c^3 m^2}\right)\left\{\frac{\omega^4}{\omega^2\gamma^2+(\omega^2-\omega_0^2)^2}\right\} \text{J s}^{-1} \tag{8.29}$$

The scattering cross-section $\sigma_s'$ for a bound charge is obtained by dividing $W_0'$ by the time averaged rate of flow of energy per unit area $W_i$ in the incoming wave. The value of $W_i$ is given by Eqn (8.26) as

$$W_i = \varepsilon_0 c E_0^2/2 \ \ \mathrm{J\,s^{-1}\,m^{-2}}$$

and therefore

$$\sigma_s' = \left(\frac{q^4}{6\pi\varepsilon_0^2 c^4 m^2}\right)\left\{\frac{\omega^4}{\omega^2\gamma^2 + (\omega^2 - \omega_0^2)^2}\right\} \tag{8.30}$$

or, using the results shown in Eqn (8.27)

$$\sigma_s' = \sigma_s\left\{\frac{\omega^4}{\omega^2\gamma^2 + (\omega^2 - \omega_0^2)^2}\right\} \tag{8.31}$$

where $\sigma_s$ is the scattering cross-section for a free charge $q$. Equation (8.31) shows that the maximum value of the scattering cross-section occurs at the resonant frequency of the bound charge when $\sigma_s' = \sigma_s(\omega_0/\gamma)^2$ m$^2$. At frequencies far from resonance if $\omega \ll \omega_0$ and $\omega\gamma \ll \omega_0^2$ the scattering cross-section becomes

$$\sigma_s' = \left(\frac{\omega}{\omega_0}\right)^4 \sigma_s \ \ \mathrm{m^2} \tag{8.32}$$

This is the result for Rayleigh scattering and is of importance in turbidity measurements which are used to estimate the concentration of particles in a solution. The general conditions given in Eqn (8.12) apply to this result and, in particular, the size of the particles causing the scattering must be small in comparison to the wavelength of the scattered radiation.

Our analysis of scattering by a bound charge is not restricted to an 'atom' or molecule; the model for the behaviour of the charge $q$ is sufficiently general that it may be applied to the behaviour of charge on small particles, for example, a biological cell held as a suspension in solution.

Notice the strong frequency dependence of the result given in Eqn (8.32). The scattering will vary with the fourth power of the frequency and so high frequencies will be most strongly scattered provided the wavelength is much greater than the diameter of the particles. If we observed the scattering of white light by small particles we would then expect that the high frequency components (blue/violet) would be more strongly scattered than the lower frequencies (red) and so the scattered radiation would appear blue. It was on this basis that Rayleigh first explained the blue colour of the sky as a consequence of atmospheric scattering. Remember that the light from the sky does not come directly from the Sun and that if there was no atmospheric scattering then the sky would be dark. Conversely, at sunset, when the light from the Sun passes through an increased thickness of atmosphere, the Sun appears red as the direct light from the Sun has lost most heavily from the blue end of the visible spectrum due to this scattering.

## 8.4 RADIATION FROM AN ELECTRIC DIPOLE ANTENNA

We now consider the radiation due to the movement of charge along a straight length of wire. If the wire is cut at its centre and connected to a sinusoidal current generator, the alternating current in the wire forms a centre fed electric dipole antenna. As usual, the electric and magnetic fields are found from the Lienard-Wiechert potentials and in this case the calculation is very similar to that given in Section 8.1 for an oscillating free charge.

Let the total length of the antenna be $l$ and arrange the coordinate axes so that the antenna lies along the $z$-axis with its centre at the origin as shown in Fig. 8.3.

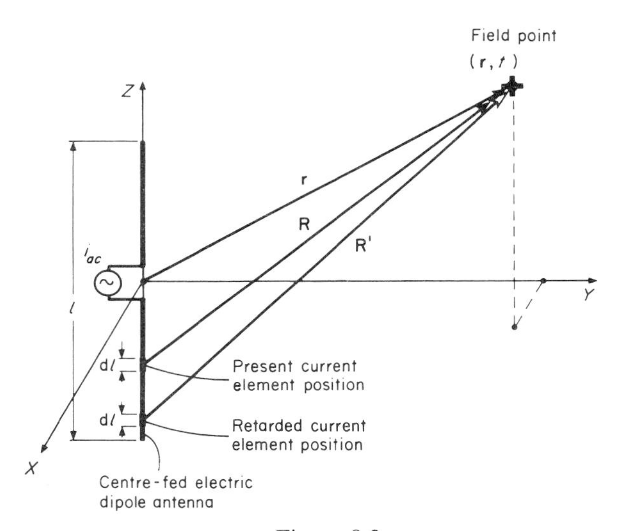


Figure 8.3

We now impose the following conditions: (i) that the velocity with which charge moves along the antenna is much less than $c$, that is,

$$v \ll c \tag{8.33}$$

(ii) that the length $l$ is very small in comparison to the distance from the antenna to the field point. Under this condition we can write $\mathbf{r} \simeq \mathbf{R} \simeq \mathbf{R}'$ and the situation is similar to that discussed in Section 8.1 where the condition was expressed in Eqn (8.12) as $R \gg \lambda$ and $L \ll \lambda$. In the present case we write the condition as

$$\boxed{\begin{gathered}\mathbf{r} \simeq \mathbf{R} \simeq \mathbf{R}' \\ R \gg \lambda \gg l\end{gathered}} \tag{8.34}$$

(iii) that at any time the current flowing in the antenna is the same at all points along its length and is given by

$$I = I_0 \sin \omega t \tag{8.35}$$

The total vector potential at a field point is found by adding the contributions from the current elements along the length of the wire. Each element of length d$l$ can be taken as carrying a charge $q$ moving with velocity $v$. The contribution from each current element is given by the Lienard-Wiechert potential

$$\mathbf{A}(\mathbf{r}, t) = \left(\frac{q}{4\pi\varepsilon_0 c^2}\right)\left\{\frac{\mathbf{v}'(\mathbf{r}_p', t')}{R'[1 - \mathbf{v}' \cdot \mathbf{n}_p'/c]}\right\}$$

The condition that the velocity of the charge is much less than $c$ means that we can write

$$\mathbf{A}(\mathbf{r}, t) = \left(\frac{q}{4\pi\varepsilon_0 c^2}\right)\left\{\frac{\mathbf{v}'(\mathbf{r}_p', t')}{R'}\right\}$$

while the condition (in Eqn (8.34)), restricting us to large distances from the antenna, means that $R \simeq R'$ and therefore

$$\mathbf{A}(\mathbf{r}, t) = \left(\frac{q}{4\pi\varepsilon_0 c^2}\right)\left\{\frac{\mathbf{v}'(\mathbf{r}_p', t')}{R}\right\}$$

The product of charge and velocity gives current and we can write

$$\begin{aligned} q\mathbf{v}'(\mathbf{r}_p', t') &= \mathbf{k}I(t')\,\mathrm{d}l \\ &= \mathbf{k}I_0 \sin[\omega(t - |\mathbf{r} - \mathbf{r}_p'|/c)]\,\mathrm{d}l \\ &= \mathbf{k}I_0 \sin[\omega(t - R'/c)]\,\mathrm{d}l \end{aligned}$$

which under the condition $R \gg \lambda \gg l$ gives

$$q\mathbf{v}'(\mathbf{r}_p', t') = \mathbf{k}I_0 \sin[\omega(t - R/c)]\,\mathrm{d}l$$

and therefore

$$\mathbf{A}(\mathbf{r}, t) = \mathbf{k}\,\frac{I_0 \sin[\omega(t - R/c)]}{4\pi\varepsilon_0 c^2 R}\,\mathrm{d}l$$

In integrating this expression over the total length of the antenna we can take $R$ as constant, since we have imposed the condition $R \gg l$ and therefore the total potential due to the antenna is given by

$$\begin{aligned} \text{Total } \mathbf{A}(\mathbf{r}, t) &= \mathbf{k}\,\frac{I_0 \sin[\omega(t - R/c)]}{4\pi\varepsilon_0 c^2 R}\,l \\ &= \mathbf{k}(I_0 l)\,\frac{\sin[\omega(t - R/c)]}{4\pi\varepsilon_0 c^2 R} \end{aligned} \tag{8.36}$$

This result is very similar to the corresponding result for a free charge given in Eqn (8.13). Comparing these two equations with Eqns (8.22) and (8.24) we can immediately write down the result for the magnetic and electric fields due to the antenna by simply replacing $(q^2E_0/m)$ by $(I_0 l\omega)$

$$\mathbf{B} = \frac{I_0 l\omega}{4\pi\varepsilon_0 c^3}\frac{\cos[\omega(t-R/c)]}{R}\left(\mathbf{k}\times\frac{\mathbf{R}}{R}\right) \tag{8.37}$$

$$\mathbf{E} = \frac{I_0 l\omega}{4\pi\varepsilon_0 c^2}\frac{\cos[\omega(t-R/c)]}{R}\left[\left(\mathbf{k}\times\frac{\mathbf{R}}{R}\right)\times\frac{\mathbf{R}}{R}\right] \tag{8.38}$$

where $R \gg \lambda \gg l$.

The results show that $B = (1/c)E$ and that the maximum value for the fields occurs (for a given value of $R$) when the angle between $\mathbf{k}$ and $\mathbf{R}/R$ is 90°, that is, in the $x$-$y$ plane. At a large distance from the antenna the fields are those for a transverse electromagnetic wave propagating radially outward from the antenna. By again comparing these results with the corresponding results for the free charge we can write down the result for the total time average rate of flow of energy from the antenna as

$$\boxed{W = \frac{I_0^2 l^2 \omega^2}{12\pi\varepsilon_0 c^3}\ \mathrm{J\ s^{-1}}} \tag{8.39}$$

by comparison with Eqn (8.25).

An alternative form of this result is

$$\boxed{W = \frac{\pi I_0^2}{3\varepsilon_0 c}\left(\frac{l}{\lambda}\right)^2\ \mathrm{J\ s^{-1}}} \tag{8.40}$$

where $R \gg \lambda \gg l$.

## 8.5 RADIATION RESISTANCE OF AN ANTENNA

The radiation resistance $R_r$ of an antenna is defined as the resistance which would dissipate energy at the same rate as the antenna when fed with the same current. The rate at which the ideal resistance $R_r$ dissipates energy when fed with a current $I = I_0 \sin \omega t$ is $\frac{1}{2}I_0^2 R_r$. Equating this with the expression for $W$ given in Eqn (8.40) gives the result

$$R_r = \frac{2\pi}{3\varepsilon_0 c}\left(\frac{l}{\lambda}\right)^2 \tag{8.41}$$

for an electric dipole antenna.

## 8.6 GAIN OF AN ANTENNA

The gain of an antenna is defined as the ratio of the total time average rate at which energy is radiated by an isotropic radiator such that it gives the same rate of radiation, *in a particular direction*, as the given antenna, to the total time average energy radiated by the antenna. The gain is a measure of the extent to which the radiation of an antenna is directional. Then the gain $g$, is given by

$$g = \frac{4\pi R^2 \bar{S}_R}{W} \tag{8.42}$$

where $\bar{S}_R$ is the magnitude of the time average Poynting vector in the direction of interest. For the electric dipole antenna the gain is a maximum in the plane containing the dipole and perpendicular to its axis (the $x$-$y$ plane in Section 8.3 above) and $g$ is given by

$$g = \frac{(4\pi R^2)\left\{\dfrac{I_0^2 l^2 \omega^2}{32\pi^2 \varepsilon_0 c^3}\left(\dfrac{1}{R^2}\right)\right\}}{\left\{\dfrac{I_0^2 l^2 \omega^2}{12\pi \varepsilon_0 c^3}\right\}}$$

that is

$$g = 3/2$$

which is the maximum gain for an electric dipole antenna.

## 8.7 RADIATION FROM A MAGNETIC DIPOLE ANTENNA

The radiation from a current loop can be treated in a similar way to that given for the electric dipole antenna in Section 8.4.

The circular loop of radius $a$ can be taken to lie in the $x$-$y$ plane with its centre at the origin as shown in Fig. 8.4. Because of the symmetry of the problem,

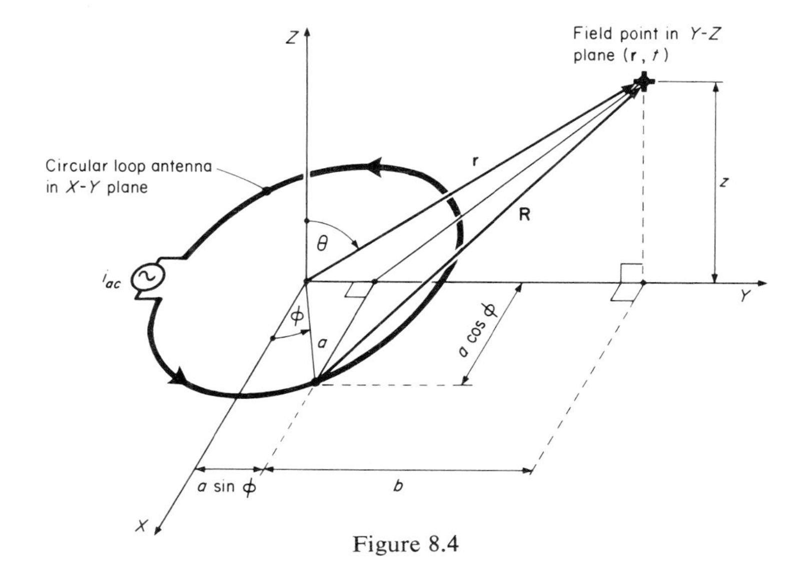


Figure 8.4

there is no loss of generality in taking the field point to lie in the $y$-$z$ plane. Care must be taken when dealing with the retarded velocity term in the Lienard-Wiechert potential, but otherwise the problem of finding the total magnetic vector potential is straightforward. This calculation is left as an exercise and the detailed working is given in Appendix 8.1. The result is that

$$\text{Total } \mathbf{A}(\mathbf{r}, t) = \mathbf{i}\,\frac{I_0 \pi a^2(\omega/c)\sin\theta\cos[\omega(t-r/c)]}{4\pi\varepsilon_0 c^2 r} \tag{8.43}$$

where $\mathbf{r} \simeq \mathbf{R}$.

This is similar to the corresponding result (given in Eqn (8.36)) for the electric dipole antenna except that $l$ is replaced by $(\pi a^2\omega/c)$ and in addition the expression contains a $\sin\theta$ term. It turns out that the electric and magnetic fields are interchanged in going from the electric to the magnetic antenna. To show this, we obtain the expressions for the electric and magnetic fields for each antenna in spherical polar coordinates. The expression for curl in spherical coordinates is given in Appendix B and the problem is simplified by taking the field point in the $y$-$z$ plane.

The results are, for the electric dipole antenna,

$$\mathbf{E}(\mathbf{r}, t) = \frac{I_0 l\omega}{4\pi\varepsilon_0 c^2}\left\{\frac{\sin\theta\cos[\omega(t-r/c)]}{r}\right\}\mathbf{i}_\theta \tag{8.44}$$

$$\mathbf{B}(\mathbf{r}, t) = \frac{I_0 l\omega}{4\pi\varepsilon_0 c^3}\left\{\frac{\sin\theta\cos[\omega(t-r/c)]}{r}\right\}\mathbf{i}_\phi \tag{8.45}$$

while for the magnetic dipole antenna,

$$\mathbf{E}(\mathbf{r}, t) = \frac{I_0 \pi a^2\omega^2}{4\pi\varepsilon_0 c^3}\left\{\frac{\sin\theta\sin[\omega(t-r/c)]}{r}\right\}\mathbf{i}_\phi \tag{8.46}$$

$$\mathbf{B}(\mathbf{r}, t) = -\frac{I_0 \pi a^2\omega^2}{4\pi\varepsilon_0 c^4}\left\{\frac{\sin\theta\sin[\omega(t-r/c)]}{r}\right\}\mathbf{i}_\theta \tag{8.47}$$

where $\mathbf{r} \simeq \mathbf{R}$.

The working necessary to obtain these results is left as an exercise and a complete solution is given in Appendix 8.2.

By comparing Eqns (8.46) and (8.47) with Eqns (8.44) and (8.45), the time average rate at which energy is radiated $W$ from the magnetic antenna can be found by reference to Eqn (8.39) which gives the corresponding result for the electric antenna. For the magnetic dipole antenna we obtain

$$W = \frac{I_0^2\omega^2}{12\pi\varepsilon_0 c^3}(\pi a^2)^2\left(\frac{\omega}{c}\right)^2 \text{ J s}^{-1} \tag{8.48}$$

The radiation resistance is readily shown to be

$$R_r = \frac{8\pi^5 a^4}{3\varepsilon_0 c\lambda^4} \tag{8.49}$$

## 8.8 LIMITATIONS IN OUR TREATMENT OF ANTENNAS

The following are some of the limitations which should be kept in mind when applying our results to antennas:

(a) We have only considered the fields at large distances from the antenna: the so-called radiation fields or far fields.

(b) Our antennas have been considered to be in free space and in practice the boundary conditions imposed by surrounding objects (the Earth, the antenna support system, nearby buildings etc.) are very important.

(c) Our results are restricted to the wavelength range where $\lambda$ is much greater than the dimensions of the antenna.

# APPENDIX 8.1

### To obtain the Vector Potential due to a Magnetic Dipole Antenna

*Solution*

The vector potential at a point in the field due to the current loop must be in a plane parallel to the plane of the loop and perpendicular to **r**. This is because the vector addition of the potentials (see Fig. 8.5) due to two elements with the same $y$-coordinates and with $x$-coordinates of opposite sign gives cancellation of the $y$-components and addition of the $x$-components.

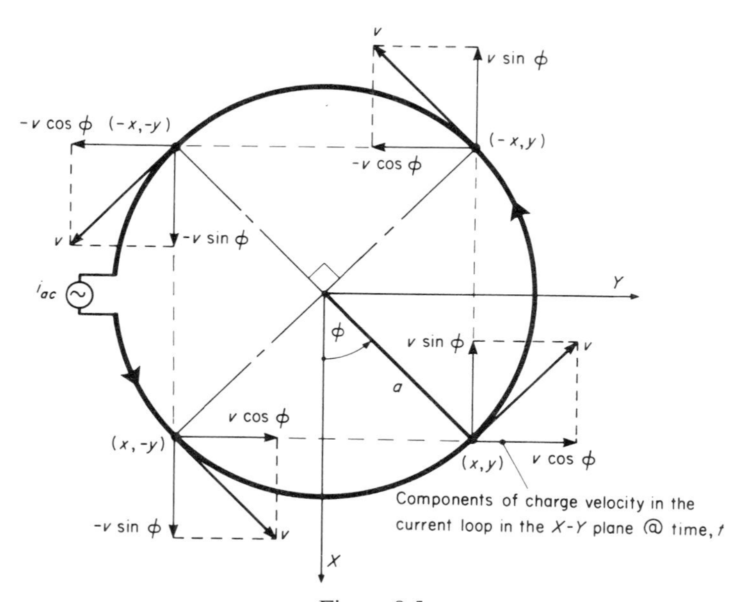


Figure 8.5

The total vector potential at ($\mathbf{r}$, $t$) due to the current loop is non-zero because the contributions from elements with the same $x$-coordinates and with $y$-coordinates of opposing sign do not cancel since they are at different distances from the point ($\mathbf{r}$, $t$). The calculation of the total vector potential $\mathbf{A}$ at ($\mathbf{r}$, $t$) is concerned with the $x$-components of the velocity terms. Applying the conditions that

$$v \ll c$$
$$R \gg \lambda \gg a$$
$$I = I_0 \sin \omega t$$

means that we can write

$$\text{Total } \mathbf{A}(\mathbf{r}, t) = \oint_{\text{loop}} \frac{q[v'(\mathbf{r}'_p, t') \sin \phi \mathbf{i}]}{4\pi\varepsilon_0 c^2 R}$$

where the distance $R$ is

$$R = \sqrt{a^2 \cos^2 \phi + z^2 + b^2}$$

as shown in Fig. 8.4. This expression for $R$ can be written as

$$R = \sqrt{a^2 \cos^2 \phi + z^2 + \{[r^2 - z^2]^{\frac{1}{2}} - a \sin \phi\}^2}$$

that is

$$R = \sqrt{a^2 + r^2 - 2a(r^2 - z^2)^{\frac{1}{2}} \sin \phi}$$

but $(a/r)^2 \ll 1$, therefore

$$R = r[1 - 2a(\sin \phi)(r^2 - z^2)^{\frac{1}{2}}/r^2]^{\frac{1}{2}}$$
$$= r[1 - a \sin \phi (r^2 - z^2)^{\frac{1}{2}}/r^2 + \cdots]$$

$$\therefore \quad R \doteqdot r - a \sin \phi \sin \theta$$

while

$$\frac{1}{R} \simeq \frac{1}{r}$$

As in the case of the electric dipole antenna, we write

$$qv'(\mathbf{r}'_p, t') = I(t')\,\mathrm{d}l$$
$$I(t') = I_0 \sin \omega t'$$
$$= I_0 \sin[\omega(t - R/c)]$$

therefore

$$\sin \omega t' = \sin\left\{\omega(t - r/c) + \frac{\omega a}{c} \sin \phi \sin \theta\right\}$$

$$= \sin[\omega(t - r/c)] \cos\left\{\frac{\omega a}{c} \sin \phi \sin \theta\right\}$$

$$+ \cos[\omega(t - r/c)] \sin\left\{\frac{\omega a}{c} \sin \phi \sin \theta\right\}$$

and since $\omega a/c \approx 0$ this is

$$\sin \omega t' = \sin[\omega(t-r/c)] + \cos[\omega(t-r/c)]\,\frac{\omega a}{c}\sin\phi\sin\theta$$

Then the total vector potential at $(\mathbf{r}, t)$ is

$$\mathbf{i}\int_0^{2\pi}\left\{\frac{I_0 \sin\phi\sin[\omega(t-r/c)]}{4\pi\varepsilon_0 c^2 r} + \frac{I_0\left(\dfrac{\omega a}{c}\right)\sin^2\phi\sin\theta\cos[\omega(t-r/c)]}{4\pi\varepsilon_0 c^2 r}\right\} a\,\mathrm{d}\phi$$

$$= \mathbf{i}\left\{0 + \frac{I_0\dfrac{\omega a^2}{c}\sin\theta\cos[\omega(t-r/c)]\pi}{4\pi\varepsilon_0 c^2 r}\right\}$$

that is

$$\text{Total } \mathbf{A}(\mathbf{r},t) = \mathbf{i}\,\frac{I_0\left(\dfrac{\omega}{c}\right)(\pi a^2)\sin\theta\cos[\omega(t-r/c)]}{4\pi\varepsilon_0 c^2 r}$$

which is the result given in Eqn (8.43).

## APPENDIX 8.2

**(i) To obtain the electric and magnetic fields, in spherical polar coordinates, at a point in the Y-Z plane, due to an electric dipole antenna located at the origin.**

*Solution*

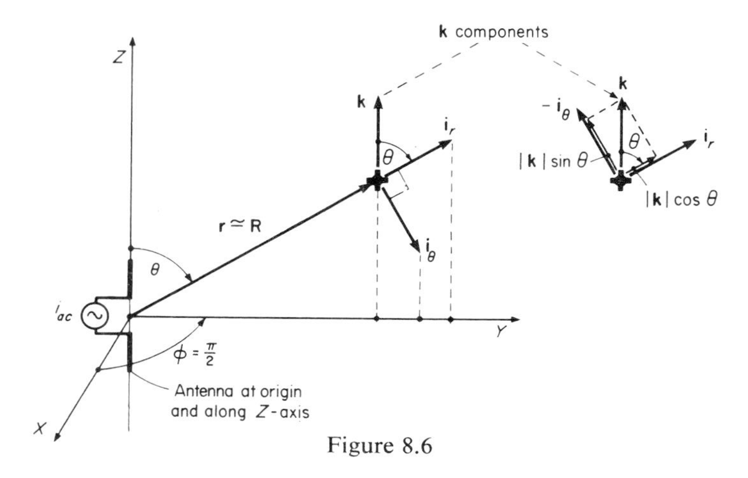


Figure 8.6

In Section 8.4 we found that the vector potential for an electric dipole antenna was (for a large distance from the antenna) given by

$$\text{Total } \mathbf{A}(\mathbf{r}, t) = \mathbf{k}\left(\frac{I_0 l}{4\pi\varepsilon_0 c^2}\right) \frac{\sin[\omega(t-r/c)]}{r}$$

For a field point in the $y$-$z$ plane, the transformation to spherical polar coordinates is simple.

The unit vector $\mathbf{k}$ is then

$$\mathbf{k} = \mathbf{i}_r \cos\theta - \mathbf{i}_\theta \sin\theta$$

and the expression for the vector potential can be written in spherical coordinates as

$$\text{Total } \mathbf{A}(\mathbf{r}, t) = \left(\frac{I_0 l}{4\pi\varepsilon_0 c^2}\right) \frac{\sin[\omega(t-r/c)]}{r} \{\mathbf{i}_r \cos\theta - \mathbf{i}_\theta \sin\theta\}$$

In spherical coordinates the curl of $\mathbf{A}$ is given (Appendix B) as

$$\nabla \times \mathbf{A} = \mathbf{i}_r\left(\frac{1}{r\sin\theta}\right)\left\{\frac{\partial}{\partial\theta}(A_\phi \sin\theta) - \frac{\partial A_\theta}{\partial\phi}\right\}$$

$$+\mathbf{i}_\theta\left(\frac{1}{r\sin\theta}\right)\left\{\frac{\partial A_r}{\partial\phi} - \sin\theta\frac{\partial}{\partial r}(rA_\phi)\right\}$$

$$+\mathbf{i}_\phi\left(\frac{1}{r}\right)\left\{\frac{\partial}{\partial r}(rA_\theta) - \frac{\partial A_r}{\partial\theta}\right\}$$

In this case we obtain

$$\nabla \times \mathbf{A} = \left(\frac{I_0 l}{4\pi\varepsilon_0 c^2}\right)\left\{\mathbf{i}_r[0] + \mathbf{i}_\theta[0]\right.$$

$$\left. +\mathbf{i}_\phi\left(\frac{1}{r}\right)\left[\sin\theta\cos[\omega(t-r/c)]\left(\frac{\omega}{c}\right) + \frac{\sin[\omega(t-r/c)]}{r}\sin\theta\right]\right\}$$

For large distances from the antenna ($r\omega/c \gg 1$) this gives the magnetic field as

$$\mathbf{B} = \mathbf{i}_\phi\left(\frac{I_0 l}{4\pi\varepsilon_0 c^2}\right)\frac{\sin\theta\cos[\omega(t-r/c)]\left(\frac{\omega}{c}\right)}{r}$$

Using the Maxwell relation (for free space)

$$c^2 \nabla \times \mathbf{B} = \frac{\partial \mathbf{E}}{\partial t}$$

we obtain

$$\nabla\times\mathbf{B} = \left(\frac{I_0 l}{4\pi\varepsilon_0 c^2}\right)\left(\frac{\omega}{c}\right)\left\{\mathbf{i}_r\left(\frac{1}{r\sin\theta}\right)\frac{\partial}{\partial\theta}\left(\frac{\sin^2\theta\cos[\omega(t-r/c)]}{r}\right)\right.$$

$$+\mathbf{i}_\theta\left(\frac{1}{r\sin\theta}\right)(-\sin\theta)\frac{\partial}{\partial r}\{\sin\theta\cos[\omega(t-r/c)]\}$$

$$\left.+\mathbf{i}_\phi\left(\frac{1}{r}\right)[0]\right\}$$

$$= \left(\frac{I_0 l}{4\pi\varepsilon_0 c^2}\right)\left(\frac{\omega}{c}\right)\left\{\mathbf{i}_r\frac{2\cos\theta\cos[\omega(t-r/c)]}{r^2}\right.$$

$$\left.+\mathbf{i}_\theta\frac{(-\sin\theta)\sin[\omega(t-r/c)]\left(\frac{\omega}{c}\right)}{r}+\mathbf{i}_\phi[0]\right\}$$

Again applying the condition that the field is at a large distance from the antenna ($r\omega/c \gg 1$) we can simplify this to

$$\nabla\times\mathbf{B} = \frac{I_0 l}{4\pi\varepsilon_0 c^2}\left(\frac{\omega}{c}\right)^2\left[-\mathbf{i}_\theta\frac{\sin\theta\sin[\omega(t-r/c)]}{r}\right]$$

Therefore

$$\mathbf{E} = \frac{I_0 l}{4\pi\varepsilon_0 c^2}(\omega)\left[\mathbf{i}_\theta\frac{\sin\theta\cos[\omega(t-r/c)]}{r}\right]$$

We conclude that in spherical polar coordinates the far fields for an electric dipole antenna are

$$\mathbf{E}(\mathbf{r}, t) = \frac{I_0 l\omega}{4\pi\varepsilon_0 c^2}\left\{\mathbf{i}_\theta\frac{\sin\theta\cos[\omega(t-r/c)]}{r}\right\}$$

$$\mathbf{B}(\mathbf{r}, t) = \frac{I_0 l\omega}{4\pi\varepsilon_0 c^3}\left\{\mathbf{i}_\phi\frac{\sin\theta\cos[\omega(t-r/c)]}{r}\right\}$$

These are the results given in Eqns (8.44) and (8.45).

**(ii) To obtain the electric and magnetic fields, in spherical polar coordinates, at a point in the Y-Z plane, due to a magnetic dipole antenna**

*Solution*

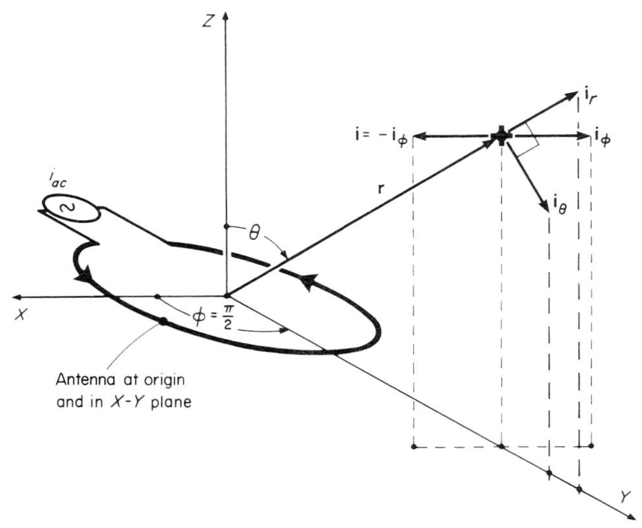


Figure 8.7

In Appendix 8.1 it was shown that, in rectangular coordinates, the vector potential at a point in the $y$-$z$ plane due to a magnetic dipole antenna located at the origin, was

$$\text{Total } \mathbf{A}(\mathbf{r}, t) = -\mathbf{i}\left(\frac{I_0 \pi a^2(\omega/c)}{4\pi\varepsilon_0 c^2}\right) \frac{\sin\theta \cos[\omega(t-r/c)]}{r}$$

Since, in this case, $\mathbf{i} = -\mathbf{i}_\phi$, the transformation to spherical coordinates is simple, giving

$$\text{Total } \mathbf{A}(\mathbf{r}, t) = \mathbf{i}_\phi\left(\frac{I_0 \pi a^2(\omega/c)}{4\pi\varepsilon_0 c^2}\right) \frac{\sin\theta \cos[\omega(t-r/c)]}{r}$$

The magnetic field is found by taking the curl of **A** which in spherical coordinates is

$$\nabla \times \mathbf{A} = \mathbf{i}_r\left(\frac{1}{r\sin\theta}\right)\left\{\frac{\partial}{\partial\theta}(A_\phi \sin\theta) - \frac{\partial A_\theta}{\partial\phi}\right\}$$

$$+\mathbf{i}_\theta\left(\frac{1}{r\sin\theta}\right)\left\{\frac{\partial A_r}{\partial\phi} - \sin\theta \frac{\partial}{\partial r}(rA_\phi)\right\}$$

$$+\mathbf{i}_\phi\left(\frac{1}{r}\right)\left\{\frac{\partial}{\partial r}(rA_\theta) - \frac{\partial A_r}{\partial\theta}\right\}$$

In this case we obtain

$$\nabla \times \mathbf{A} = \left(\frac{I_0 \pi a^2(\omega/c)}{4\pi\varepsilon_0 c^2}\right)\left\{\mathbf{i}_r\left(\frac{1}{r \sin\theta}\right)\left[\frac{\partial}{\partial\theta}\left(\frac{\sin^2\theta \cos[\omega(t-r/c)]}{r}\right)\right]\right.$$

$$+\mathbf{i}_\theta\left(\frac{1}{r \sin\theta}\right)\left[0-\sin\theta \frac{\partial}{\partial r}\{\sin\theta \cos[\omega(t-r/c)]\}\right]$$

$$\left.+\mathbf{i}_\phi\left(\frac{1}{r}\right)[0]\right\}$$

$$= \left(\frac{I_0 \pi a^2(\omega/c)}{4\pi\varepsilon_0 c^2}\right)\left\{\mathbf{i}_r \frac{2\cos\theta \cos[\omega(t-r/c)]}{r^2}\right.$$

$$\left.+\mathbf{i}_\theta \frac{(-\sin\theta)\sin[\omega(t-r/c)]\left(\frac{\omega}{c}\right)}{r} + \mathbf{i}_\phi[0]\right\}$$

For large distances from the antenna ($r\omega/c \gg 1$) this reduces to

$$\mathbf{B}(\mathbf{r}, t) = -\left(\frac{I_0 \pi a^2(\omega/c)^2}{4\pi\varepsilon_0 c^2}\right)\frac{\sin\theta \sin[\omega(t-r/c)]}{r}\,\mathbf{i}_\theta$$

The electric field is then found from the Maxwell relation (free space)

$$c^2 \nabla \times \mathbf{B} = \frac{\partial \mathbf{E}}{\partial t}$$

In this case

$$\nabla \times \mathbf{B} = -\left(\frac{I_0 \pi a^2(\omega/c)^2}{4\pi\varepsilon_0 c^2}\right)\left\{\mathbf{i}_r\left(\frac{1}{r\sin\theta}\right)[0]\right.$$

$$\left.+\mathbf{i}_\theta\left(\frac{1}{r\sin\theta}\right)[0] + \mathbf{i}_\phi\left(\frac{1}{r}\right)\frac{\partial}{\partial r}\{\sin\theta \sin[\omega(t-r/c)]\}\right\}$$

$$= -\left(\frac{I_0 \pi a^2(\omega/c)^2}{4\pi\varepsilon_0 c^2}\right)\left\{\mathbf{i}_\phi \frac{\sin\theta \cos[\omega(t-r/c)](-\omega/c)}{r}\right\}$$

Hence, the electric field is

$$\mathbf{E}(\mathbf{r}, t) = \frac{I_0 \pi a^2 \omega^2}{4\pi\varepsilon_0 c^2}\left\{\mathbf{i}_\phi \frac{\sin\theta \sin[\omega(t-r/c)](1/c)}{r}\right\}$$

We conclude that for a magnetic dipole antenna the far fields are given by

$$\mathbf{E}(\mathbf{r}, t) = \left(\frac{I_0 \pi a^2 \omega^2}{4\pi\varepsilon_0 c^3}\right)\left\{\frac{\sin\theta \sin[\omega(t-r/c)]}{r}\right\}\mathbf{i}_\phi$$

$$\mathbf{B}(\mathbf{r}, t) = -\left(\frac{I_0 \pi a^2 \omega^2}{4\pi\varepsilon_0 c^4}\right)\left\{\frac{\sin\theta \sin[\omega(t-r/c)]}{r}\right\}\mathbf{i}_\theta$$

These are the results stated in Eqns (8.46) and (8.47).

## PROBLEMS

1.
Use the Thomson formula to calculate the total scattering cross-section for (a) a free electron, (b) a free proton.

2.
The refractive index of carbon dioxide gas (at 20 °C and 1 atmosphere) is approximately 1.00045 when measured using sodium light ($\lambda$ = 5892 Å). Remembering that any gas contains Avogadro's number of molecules in 22.4 $m^3$ at 0 °C and at 1 atmosphere, use Eqns (4.27) and (8.30) to obtain a value for the scattering cross-section of carbon dioxide.

3.
We have discussed the scattering of radiation in the visible spectrum. Given that violet light has a wavelength of 4000 Å, and red light has a wavelength of 7200 Å, calculate the factor by which the scattering of the violet light is increased over the scattering of red light. Assume that the particles causing the scattering have a diameter much less than 4000 Å.

4.
Estimate the current that must be supplied to radiate 1 kW at 1 MHz from an electric dipole antenna of length equal to one tenth of a wavelength.

5.
Calculate the maximum value of the electric field at a distance of 10 km from the antenna described in Problem 4.

6.
(a)
Consider a circular loop of wire of radius $a$ (m), lying in the $x$-$y$ plane and carrying a constant current $I_0$ (A). If the current is flowing clockwise when viewed along the $z$-axis in the direction of $z$ increasing, show that the magnetic field at a point in the $y$-$z$ plane and distant from the loop, is given by

$$\mathbf{B}(\mathbf{r}) = \frac{I_0 a^2 \pi}{4\pi\varepsilon_0 c^2}\left\{\mathbf{i}_r \frac{2\cos\theta}{r^3} + \mathbf{i}_\theta \frac{\sin\theta}{r^3}\right\}$$

*Note:* the working is essentially similar to that given in Chapter 8 for the magnetic dipole antenna.

(b)
Now show that if we define the magnetic dipole moment $\mathbf{m}$ of the current loop as

$$\mathbf{m} = I_0 a^2 \pi \mathbf{k}$$

then the above expression for the magnetic field is equivalent to

$$\mathbf{B}(\mathbf{r}) = \frac{1}{4\pi\varepsilon_0 c^2}\left\{\frac{3(\mathbf{m}\cdot\mathbf{r})\mathbf{r}}{r^5} - \frac{\mathbf{m}}{r^3}\right\}$$

This result is analogous to the result given in Exercise 1.2 for the electric field due to an electric dipole of moment $\mathbf{p}$.

## FURTHER READING

Scattering is treated by Stratton and by Panofsky & Philips; for an advanced and detailed treatment (including applications) see Kerker. See (for example), Marion, Stratton, Paris & Hurd for antennas. Practical design details of antenna systems are given by Jordan & Balmain and by Williams.

## Chapter 9

# POLAR MATERIALS

So far we have given little attention to the important materials where the molecules have permanent electric dipoles. The permanent separation of charge within a molecule means that a calculation of the internal field within a material becomes very complicated. Even in gases and liquids where the properties are isotropic, the local field cannot be properly treated in terms of the average position of the dipoles. It is this basic problem which underlies the difficulties in attempting to provide a simple treatment of ferroelectric materials from a molecular point of view. There are similar difficulties with ferromagnetic materials. A fuller discussion of this point is given later. Despite these problems with materials in which the molecules have a permanent electric dipole, we can obtain a very general and useful result on the basis of our equation for the conservation of electromagnetic energy and some elementary thermodynamics.

## 9.1 ELECTROMAGNETIC ENERGY IN A DIELECTRIC MATERIAL

In Chapter 3 we found that for an isotropic material we may write the Maxwell equations as

$$\nabla\cdot(\mathbf{E}+\mathbf{P}/\varepsilon_0) = \rho_f/\varepsilon_0$$

$$\nabla\times\mathbf{E} = -\partial\mathbf{B}/\partial t$$

$$\nabla\cdot\mathbf{B} = 0$$

$$c^2\nabla\times\mathbf{B} = \mathbf{J}_f/\varepsilon_0+\frac{\partial}{\partial t}(\mathbf{E}+\mathbf{P}/\varepsilon_0)$$

If these equations are analysed by the method used in Chapter 1, Section 1.7, we obtain the energy conservation equation

$$\varepsilon_0 c^2\nabla\cdot\mathbf{E}\times\mathbf{B}+\mathbf{E}\cdot\mathbf{J}_f+\varepsilon_0\mathbf{E}\cdot\frac{\partial}{\partial t}(\mathbf{E}+\mathbf{P}/\varepsilon_0)+\varepsilon_0 c^2\mathbf{B}\cdot\frac{\partial\mathbf{B}}{\partial t} = 0$$

and each term represents the rate of energy flow per unit volume of the material. The term $\varepsilon_0\mathbf{E}\cdot\partial/\partial t(\mathbf{E}+\mathbf{P}/\varepsilon_0)$ is associated with the electric field *only* and we can identify the term $\varepsilon_0\mathbf{E}\cdot\mathrm{d}(\mathbf{E}+\mathbf{P}/\varepsilon_0)$ as the change in energy per unit volume of material due to an infinitesimal change d**E** in the electric field. From our definition of the relative permittivity $\varepsilon$ we can write this term as $\varepsilon_0 \mathbf{E} \cdot \mathrm{d}(\varepsilon \mathbf{E})$ for an isotropic dielectric.

Now let us take a look at the entropy changes which occur in a material as we change only two parameters: the electric field **E** and the temperature $T$. The definition of an entropy change $\mathrm{d}\mathscr{S}$ is given by

$$\mathrm{d}\mathscr{S} = \mathrm{d}Q/T$$

where $\mathrm{d}Q$ is the amount of heat put into the system at the temperature $T$. If we write $U$ for the energy per unit volume of the material, then the principle of conservation of energy requires that, under the conditions given above

$$\begin{aligned} \mathrm{d}U &= \mathrm{d}Q + \varepsilon_0 E \, \mathrm{d}(\varepsilon E) \\ &= \mathrm{d}Q + \varepsilon_0 E(E \, \mathrm{d}\varepsilon + \varepsilon \, \mathrm{d}E) \end{aligned}$$

which is

$$\mathrm{d}U = \mathrm{d}Q + E^2 \varepsilon_0 \, \mathrm{d}\varepsilon + \frac{\varepsilon_0 \varepsilon}{2} \, \mathrm{d}(E^2)$$

But the energy of the material is a function of the temperature and the electric field only, and so we can write

$$U = U(E^2, T)$$

and then

$$\mathrm{d}U = \frac{\partial U}{\partial (E^2)} \, \mathrm{d}(E^2) + \frac{\partial U}{\partial T} \, \mathrm{d}T$$

where we are taking $\varepsilon$ to be a function of $T$. Equating the two expressions for $\mathrm{d}U$ gives

$$\frac{\partial U}{\partial (E^2)} \, \mathrm{d}(E^2) + \frac{\partial U}{\partial T} \, \mathrm{d}T = \mathrm{d}Q + E^2 \varepsilon_0 \, \mathrm{d}\varepsilon + \frac{\varepsilon_0 \varepsilon}{2} \, \mathrm{d}(E^2)$$

that is

$$\mathrm{d}Q = \frac{\partial U}{\partial (E^2)} \, \mathrm{d}(E^2) + \frac{\partial U}{\partial T} \, \mathrm{d}T - E^2 \varepsilon_0 \frac{\mathrm{d}\varepsilon}{\mathrm{d}T} \, \mathrm{d}T - \frac{\varepsilon_0 \varepsilon}{2} \, \mathrm{d}(E^2)$$

Dividing across by $T$ gives the expression for $\mathrm{d}\mathscr{S}$ and the terms may then be grouped to give

$$\mathrm{d}\mathscr{S} = \left( \frac{1}{T} \frac{\partial U}{\partial T} - \frac{\varepsilon_0 E^2}{T} \frac{\mathrm{d}\varepsilon}{\mathrm{d}T} \right) \mathrm{d}T + \left( \frac{1}{T} \frac{\partial U}{\partial (E^2)} - \frac{\varepsilon_0 \varepsilon}{2T} \right) \mathrm{d}(E^2) \tag{9.1}$$

Since the entropy is a function of $E^2$ and $T$ only

$$\mathscr{S} = \mathscr{S}(E^2, T)$$

and since $\mathscr{S}$ is taken to be a mathematically well behaved function

$$\mathrm{d}\mathscr{S} = \frac{\partial \mathscr{S}}{\partial T}\,\mathrm{d}T + \frac{\partial \mathscr{S}}{\partial (E^2)}\,\mathrm{d}(E^2) \tag{9.2}$$

and

$$\frac{\partial}{\partial (E^2)}\left(\frac{\partial \mathscr{S}}{\partial T}\right) = \frac{\partial}{\partial T}\left(\frac{\partial \mathscr{S}}{\partial (E^2)}\right) \tag{9.3}$$

The expressions for $\partial\mathscr{S}/\partial T$ and $\partial\mathscr{S}/\partial(E^2)$ can be identified by comparing Eqns (9.1) and (9.2); then from Eqn (9.3) we can write

$$\frac{\partial}{\partial (E^2)}\left(\frac{1}{T}\frac{\partial U}{\partial T} - \frac{\varepsilon_0 E^2}{T}\frac{\mathrm{d}\varepsilon}{\mathrm{d}T}\right) = \frac{\partial}{\partial T}\left(\frac{1}{T}\frac{\partial U}{\partial (E^2)} - \frac{\varepsilon_0 \varepsilon}{2T}\right)$$

that is

$$\frac{1}{T}\frac{\partial^2 U}{\partial (E^2)\,\partial T} - \frac{\varepsilon_0}{T}\frac{\mathrm{d}\varepsilon}{\mathrm{d}T} = \frac{1}{T}\frac{\partial^2 U}{\partial T\,\partial (E^2)} - \frac{1}{T^2}\frac{\partial U}{\partial (E^2)} + \frac{\varepsilon_0 \varepsilon}{2T^2} - \frac{\varepsilon_0}{2T}\frac{\mathrm{d}\varepsilon}{\mathrm{d}T}$$

which gives

$$\frac{\partial U}{\partial (E^2)} = \frac{\varepsilon_0 T}{2}\frac{\mathrm{d}\varepsilon}{\mathrm{d}T} + \frac{\varepsilon_0 \varepsilon}{2} \tag{9.4}$$

By comparing Eqns (9.1) and (9.2) we have

$$\frac{\partial \mathscr{S}}{\partial (E^2)} = \frac{1}{T}\frac{\partial U}{\partial (E^2)} - \frac{\varepsilon_0 \varepsilon}{2T}$$

which from Eqn (9.4) becomes

$$\frac{\partial \mathscr{S}}{\partial (E^2)} = \frac{\varepsilon_0}{2}\frac{\mathrm{d}\varepsilon}{\mathrm{d}T} + \frac{\varepsilon_0 \varepsilon}{2T} - \frac{\varepsilon_0 \varepsilon}{2T}$$

that is

$$\frac{\partial \mathscr{S}}{\partial (E^2)} = \frac{\varepsilon_0}{2}\frac{\mathrm{d}\varepsilon}{\mathrm{d}T}$$

Integrating partially with respect to $E^2$ gives

$$\boxed{\mathscr{S} = \frac{\varepsilon_0}{2}\frac{\mathrm{d}\varepsilon}{\mathrm{d}T}(E^2) + \mathscr{S}_0(T)} \tag{9.5}$$

where $\mathscr{S}_0(T)$ is a contribution to the entropy which depends only on the temperature $T$.

This very general result tells us that if we apply an electric field to a dielectric material, then the entropy either increases or decreases depending on whether $\mathrm{d}\varepsilon/\mathrm{d}T$ is positive or negative. That is, if $\mathrm{d}\varepsilon/\mathrm{d}T$ is positive then the application of the field serves to increase $\mathscr{S}$ which corresponds to reducing the degree of *order* within the material. Conversely, if $\mathrm{d}\varepsilon/\mathrm{d}T$ is negative, then the application of an electric field decreases the entropy and so corresponds to increasing the ordered nature of the material. A change in the sign of $d\varepsilon/dT$ will correspond to an order/disorder transition of the type which occurs when a material passes from the solid to the liquid or gas state. For solids, where the molecular dipoles are ordered into the crystal structure, we can expect that $d\varepsilon/dT$ will be positive whereas for liquids or gases, where such a highly ordered state does not occur, we can expect that $d\varepsilon/dT$ will be negative. This is what is observed in practice and we now consider the physical explanation of these results.

We can understand the results of Eqn (9.5) on an intuitive basis. The polarization **P** of a polar material will be dominated by the orientational polarization of the molecular dipoles; in other words, the biggest contribution to the movement of charge within the material will be made by the swinging of the molecules to align themselves with the direction of the electric field. We are assuming here that the field is either a static or low frequency field so that the molecular movement can follow the field. Even though the molecules are free to rotate, perfect alignment is not achieved in the liquid or gas states because the thermal energy of the molecules tends to randomize their orientations. Increasing the temperature increases these disruptive effects. These are the two opposing effects, the electric field tending to pull the molecules into its direction, so increasing the values of **P** and $\varepsilon$, and the temperature which disrupts the alignment. Then, in liquids and gases, $\varepsilon$ must *decrease* with increasing temperature, that is, $d\varepsilon/dT$ will be negative.

In the solid state the situation is different. The molecules occupy essentially fixed positions about which they can oscillate slightly with thermal energy. Melting corresponds to the temperature at which the thermal energy has increased the freedom of the molecules to move to the extent that they break loose from their fixed positions. In the solid state the effect of increasing the temperature is to increase the freedom of the molecules to align with the field. Then $\varepsilon$ must increase with temperature and $d\varepsilon/dT$ is positive for a solid. The results are shown schematically in Fig. 9.1.

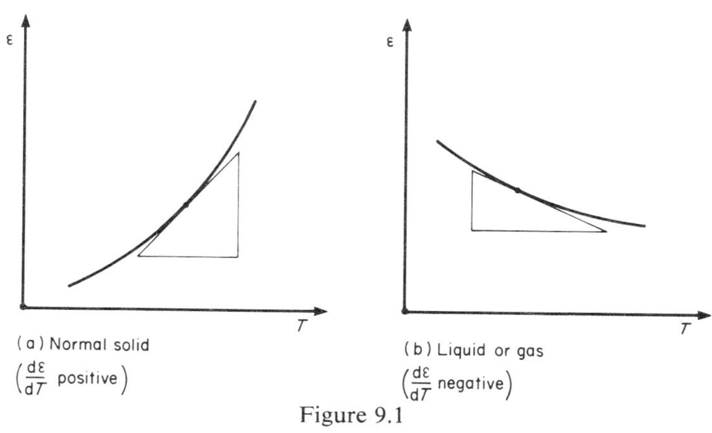


Figure 9.1

## 9.2 SOLIDS THAT ARE NOT ALWAYS SOLID

Equation (9.2) proves to be very useful in the study of solids where a molecular rearrangement can occur in the solid state. In some materials which have a very smooth molecular shape, for example, approximately spherical or cylindrical shapes, the molecules can gain rotational freedom in the solid state *before* they gain translational freedom and become liquids. When the 'rotational melting point' occurs the molecular dipoles are free to align with the applied field in much the same way as in the liquid state and so $d\varepsilon/dT$ is negative. The solid may still look quite 'solid' and may not melt until the temperature is very much higher. Examples of materials showing this kind of behaviour are the fatty esters (approximately cylindrical) where ethyl stearate, $C_{17}H_{35}CO_2C_2H_5$, has a rotational transition at +23 °C and melts at +34 °C. Hydrogen chloride is a small and approximately spherical molecule with a rotational transition at −174 °C and a melting point at −111 °C. An example of a much larger spherical molecule is camphor, $C_{10}H_{16}O$, which has a rotational transition at −35 °C and a melting point at 180 °C. The transition in camphor is shown in Fig. 9.2 where the capacitance of a parallel plate capacitor is plotted against temperature.

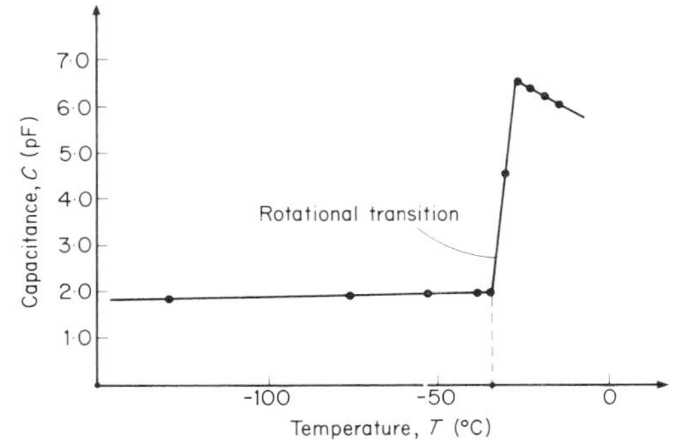


Figure 9.2 Results taken from V. Rossiter, *J. Phys. C; Solid State Phys.*, 1972, vol. 5 1969–1975.

The large and sudden increase in ε during the transition is typical of many of these materials. It shows that all of the molecules have suddenly become free to rotate. This kind of cooperative effect shows the importance of the interaction of one molecule with its neighbours through internal field effects.

## 9.3 FERROELECTRIC AND FERROMAGNETIC MATERIALS

These materials are characterized by a permanent self induced polarization below a critical temperature, usually called the Curie temperature $T_C$. Below this temperature the thermal energy which tends to randomize the orientation of the molecular dipoles (electric or magnetic) is reduced to the extent that the local fields of the dipoles produce alignment and give large scale polarization of the material. In real materials the phenomena are complicated by, for example, the occurrence of multiple ferroelectric transitions and the presence of domain structure which means that in ferromagnetic materials the spontaneous polarization occurs in different directions in different regions (or domains) within the material so that no large net magnetic field is produced outside the material. These transitions occur sharply in small temperature intervals and we can say that 'cooperative effects' are again important. Simple theories can be proposed which attempt to explain both ferroelectric and ferromagnetic behaviour in terms of a local field described by an equation of the type

$$\mathbf{E}_{\text{local}} = \mathbf{E} + \beta\mathbf{P}$$

or

$$\mathbf{B}_{\text{local}} = \mathbf{B} + \beta\mathbf{M}$$

where **M** is the magnetization per unit volume (analogous to **P**) and $\beta$ is a constant which describes the contribution to the local field by the surrounding dipoles. Such equations can lead to the prediction of spontaneous polarization and to a relation of the Curie-Weiss type describing the behaviour of the

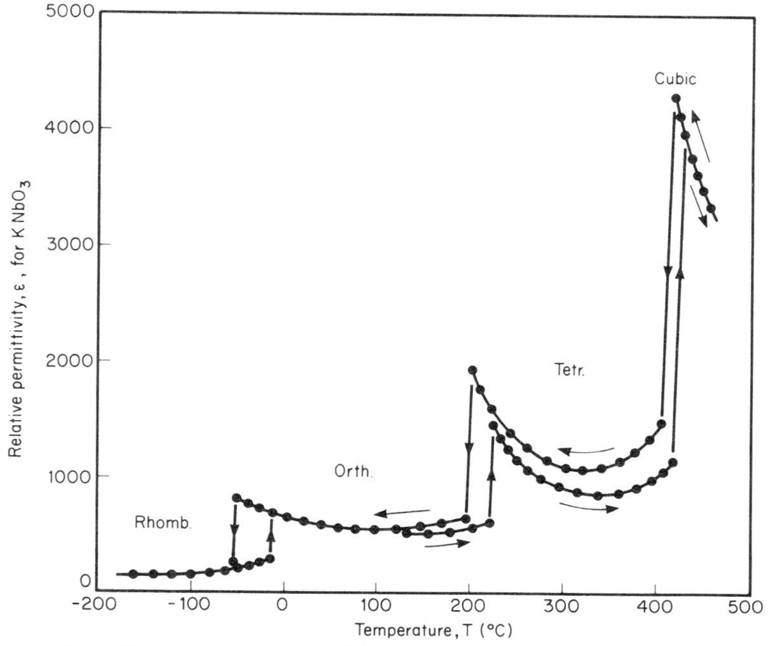


Figure 9.3 Results obtained by Shirane et al., *Phys Rev.*, **93**, 672–673 (1954).

materials above the Curie temperature. For ferroelectric materials the Curie-Weiss relationship gives the temperature dependence of the electric susceptibility $\chi$ as

$$\chi = \frac{C}{T-T_c} \text{ for } T > T_c$$

where we define $\chi$ as

$$\chi = \varepsilon - 1$$

and $C$ is a constant, $T$ is the absolute temperature, $T_C$ is the Curie temperature. Unfortunately, when the values of $\beta$ are calculated for this simple theory, they cannot be reconciled with what we know about such materials. Our earlier discussion of the local field and the derivation of the Lorentz expression do not encourage us to accept expressions of the type above for $\mathbf{E}_{\text{local}}$ and $\mathbf{B}_{\text{local}}$ in materials with a complicated structure. It is not sufficient to deal with an 'average' local field due to the average arrangement of neighbours, particularly since we know that cooperative effects are important and what happens locally may be sufficient to trigger off a chain reaction throughout the material.

The ferroelectric transitions in potassium niobate, $KNbO_3$, are shown in Fig. 9.3 to indicate the complexities of real materials. The changes in crystal structure which occur at each ferroelectric transition temperature are shown in the figure.

## FURTHER READING

Further information on polar materials is given by Feynman *et al.*, Anderson, Dekker, Fröhlich, Smyth and Hill *et al.* Detailed information on materials which are important in practical devices is given by Jaffe et al.

Appendix A

# REFERENCES TO SOURCES OF FURTHER READING

This list of references contains some of the many books dealing with electromagnetism in one form or another. The titles to which specific references have been made are indicated with a plus (+) and texts of a more advanced nature are indicated with an asterisk (*).


+J. C. Anderson, *Dielectrics*, Chapman & Hall (1967).
+J. C. Anderson, *Magnetism and Magnetic Materials*, Chapman & Hall (1968).
+C. L. Andrews, *Optics of the Electromagnetic Spectrum*, Prentice-Hall (1960).
T. G. Barnes, *Foundations of Electricity and Magnetism*, Heath (1965).
R. M. Bozorth, *Ferromagnetism*, D. Van Nostrand (1951).
+P. C. Clemmow, *An Introduction to Electromagnetic Theory*, Cambridge University Press (1973).
D. M. Cook, *The Theory of the Electromagnetic Field*, Prentice-Hall (1975).
D. R. Corson and P. Lorrain, *Introduction to Electromagnetic Fields and Waves*, Freeman (1962).
J. B. Davies and D. E. Radley, *Electromagnetic Theory* (vols 1 and 2), Oliver & Boyd (1972).
+A. J. Dekker, *Solid State Physics*, Macmillan (1970).
L. Eyges, *The Classical Electromagnetic Field*, Addison-Wesley (1972).*
+R. P. Feynman, R. B. Leighton and M. Sands, *The Feynman Lectures on Physics* (vol. 2), Addison-Wesley (1964).
+H. Fröhlich, *Theory of Dielectrics*, Clarendon Press (1958).*
R. H. Good and T. J. Nelson, *Classical Theory of Electric and Magnetic Fields*, Academic Press (1971).*
I. S. Grant and W. R. Philips, *Electromagnetism*, Wiley (1975).
W. H. Hayt, *Engineering Electromagnetics*, McGraw-Hill (1974).
+N. Hill, A. H. Price, W. E. Vaughan and M. Davies, *Dielectric Properties and Molecular Behaviour*, Van Nostrand Reinhold (1969).
+C. A. Holt, *Introduction to Electromagnetic Fields and Waves*, Wiley (1963).
J. D. Jackson, *Classical Electrodynamics*, Wiley (1962).*
+B. Jaffe, W. R. Cook and H. Jaffe, *Piezoelectric Ceramics*, Academic Press (1971).
+M. Javid and P. M. Brown, *Field Analysis and Electromagnetics*, McGraw-Hill (1963).*
F. A. Jenkins and H. E. White, *Fundamentals of Optics*, McGraw-Hill (3rd Edn, 1957).
C. T. A. Johnk, *Engineering Electromagnetic Fields and Waves*, Wiley (1975).
D. S. Jones, *The Theory of Electromagnetism*, Pergamon Press (1964).*

+E. C. Jordan and K. G. Balmain, *Electromagnetic Waves and Radiating Systems*, Prentice-Hall (2nd Edn, 1968).*
+M. Kerker, *The Scattering of Light and other Electromagnetic Radiation*, Academic Press (1969).*
+M. Kerker (ed.), *Electromagnetic Scattering*, Pergamon Press (1963).*
A. F. Kip, *Fundamentals of Electricity and Magnetism*, McGraw-Hill (1969).
+L. D. Landau and E. M. Lifshitz, *The Classical Theory of Fields*, Pergamon Press (1971).
R. V. Langmuir, *Electromagnetic Fields and Waves*, McGraw-Hill (1961).
+L. M. Magid, *Electromagnetic Fields, Energy and Waves*, Wiley (1972).
+J. B. Marion, *Classical Electromagnetic Radiation*, Academic Press (1965).
+W. K. H. Panofsky and M. Philips, *Classical Electricity and Magnetism*, Addison-Wesley (1962).*
+C. H. Papas, *Theory of Electromagnetic Wave Propagation*, McGraw-Hill (1965).
+D. T. Paris and F. K. Hurd, *Basic Electromagnetic Theory*, McGraw-Hill (1969).*
E. M. Pugh and E. W. Pugh, *Principles of Electricity and Magnetism*, Addison-Wesley (1970).
S. Ramo, J. R. Whinnery and T. van Duzer, *Fields and Waves in Communication Electronics*, Wiley (1965).
N. N. Rao, *Basic Electromagnetics with Applications*, Prentice-Hall (1972).
+J. R. Reitz and F. J. Milford, *Foundations of Electromagnetic Theory*, Addison-Wesley (1967).
F. N. H. Robinson, *Macroscopic Electromagnetism*, Pergamon Press (1973).
V. Rojansky, *Electromagnetic Fields and Waves*, Prentice-Hall (1971).
S. A. Schelkunoff, *Electromagnetic Fields*, Blaisdell (1963).
W. M. Schwarz, *Intermediate Electromagnetic Theory*, Wiley (1964).
+C. P. Smyth, *Dielectric Behaviour and Structure*, McGraw-Hill (1955).
W. R. Smythe, *Static and Dynamic Electricity*, McGraw-Hill (1968).
L. Solymar, *Lectures on Electromagnetic Theory*, Oxford University Press (1976).
A. Sommerfeld, *Electrodynamics*, Academic Press (1964).
+J. A. Stratton, *Electromagnetic Theory*, McGraw-Hill (1941).
+D. W. Tenquist, R. M. Whittle and J. Yarwood, *University Optics* (vols 1 and 2), Iliffe Books (1970).
N. Tralli, *Classical Electromagnetic Theory*, McGraw-Hill (1963).
+R. A. Waldron, *Theory of Guided Electromagnetic Waves*, Van Nostrand Reinhold (1970).*
W. L. Weeks, *Electromagnetic Theory for Engineering Applications*, Wiley (1964).
+H. P. Williams, *Antenna Theory & Design* (vol. II), Pitman (1966).*
A. A. Zaky and R. Hawley, *Fundamentals of Electromagnetic Field Theory*, Harrap (1974).

## Appendix B

# SOME USEFUL MATHEMATICAL RESULTS

## 1(a) VECTOR RELATIONSHIPS

The vectors **i**, **j**, **k** are each of unit length and their directions are in the directions of the positive $x$, $y$, $z$ axes respectively.

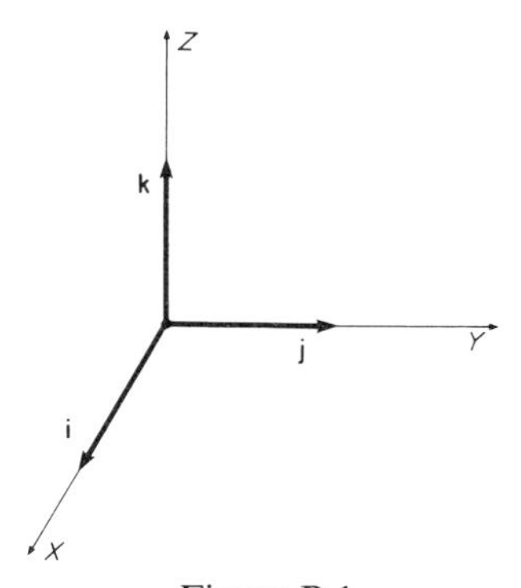


Figure B.1

Then any vector can be described in terms of these unit vectors. For example, we could write

$$\mathbf{a} = \mathbf{i}a_x + \mathbf{j}a_y + \mathbf{k}a_z$$

and the magnitude of **a** is given by

$$a = \sqrt{a_x^2 + a_y^2 + a_z^2}$$

The vector operations of addition and subtraction are such that, if

$$\mathbf{b} = \mathbf{i}b_x + \mathbf{j}b_y + \mathbf{k}b_z$$

then

$$\mathbf{a} + \mathbf{b} = \mathbf{i}(a_x + b_x) + \mathbf{j}(a_y + b_y) + \mathbf{k}(a_z + b_z)$$

and

$$\mathbf{a} - \mathbf{b} = \mathbf{i}(a_x - b_x) + \mathbf{j}(a_y - b_y) + \mathbf{k}(a_z - b_z)$$

The scalar multiplication of vectors is written

$$\begin{aligned}\mathbf{a}\cdot\mathbf{b} &= a_x b_x + a_y b_y + a_z b_z \\ &= ab\cos\theta\end{aligned}$$

where $\theta$ is the angle between **a** and **b**.

Vector multiplication is written

$$\begin{aligned}\mathbf{a}\times\mathbf{b} &= \begin{vmatrix} \mathbf{i} & \mathbf{j} & \mathbf{k} \\ a_x & a_y & a_z \\ b_x & b_y & b_z \end{vmatrix} \\ &= \mathbf{i}(a_y b_z - a_z b_y) - \mathbf{j}(a_x b_z - a_z b_x) + \mathbf{k}(a_x b_y - a_y b_x) \\ &= \mathbf{C}\text{ (say)}\end{aligned}$$

and the magnitude of the vector product is such that

$$C = |\mathbf{a}\times\mathbf{b}| = ab\sin\theta$$

where $\theta$ is the angle between **a** and **b**.

The vector **C** is perpendicular to both **a** and **b**, for example

$$\begin{aligned}\mathbf{i}\times\mathbf{j} &= \begin{vmatrix} \mathbf{i} & \mathbf{j} & \mathbf{k} \\ 1 & 0 & 0 \\ 0 & 1 & 0 \end{vmatrix} \\ &= \mathbf{k}\end{aligned}$$

The vector differential operator $\nabla$ (del) is, in rectangular coordinates

$$\nabla \equiv \mathbf{i}\frac{\partial}{\partial x} + \mathbf{j}\frac{\partial}{\partial y} + \mathbf{k}\frac{\partial}{\partial z}$$

and this operator undergoes scalar and vector multiplication like the general vectors **a** and **b** described above. Multiplication with a scalar term $u$ (say) gives the gradient of $u$ (grad $u$)

$$\nabla u = \mathbf{i}\frac{\partial u}{\partial x} + \mathbf{j}\frac{\partial u}{\partial y} + \mathbf{k}\frac{\partial u}{\partial z}$$

The divergence of a vector **a** (div **a**) is given by

$$\nabla\cdot\mathbf{a} = \frac{\partial a_x}{\partial x} + \frac{\partial a_y}{\partial y} + \frac{\partial a_z}{\partial z}$$

The curl of a vector **a** (curl **a**) is given by

$$\nabla\times\mathbf{a} = \begin{vmatrix} \mathbf{i} & \mathbf{j} & \mathbf{k} \\ \dfrac{\partial}{\partial x} & \dfrac{\partial}{\partial y} & \dfrac{\partial}{\partial z} \\ a_x & a_y & a_z \end{vmatrix}$$

$$= \mathbf{i}\left(\frac{\partial a_z}{\partial y}-\frac{\partial a_y}{\partial z}\right)-\mathbf{j}\left(\frac{\partial a_z}{\partial x}-\frac{\partial a_x}{\partial z}\right)+\mathbf{k}\left(\frac{\partial a_y}{\partial x}-\frac{\partial a_x}{\partial y}\right)$$

Scalar multiplication of the operator with itself gives

$$\nabla\cdot\nabla = \frac{\partial^2}{\partial x^2}+\frac{\partial^2}{\partial y^2}+\frac{\partial^2}{\partial z^2}$$

$$= \nabla^2$$

In spherical polar coordinates the vector **a** is described by

$$\mathbf{a} = \mathbf{i}_r a_r+\mathbf{i}_\theta a_\theta+\mathbf{i}_\phi a_\phi$$

where $\mathbf{i}_r$ is the unit vector in the direction of $r$ (only) increasing, where $\mathbf{i}_\theta$ is the unit vector in the direction of $\theta$ (only) increasing, and where $\mathbf{i}_\phi$ is the unit vector in the direction of $\phi$ (only) increasing.

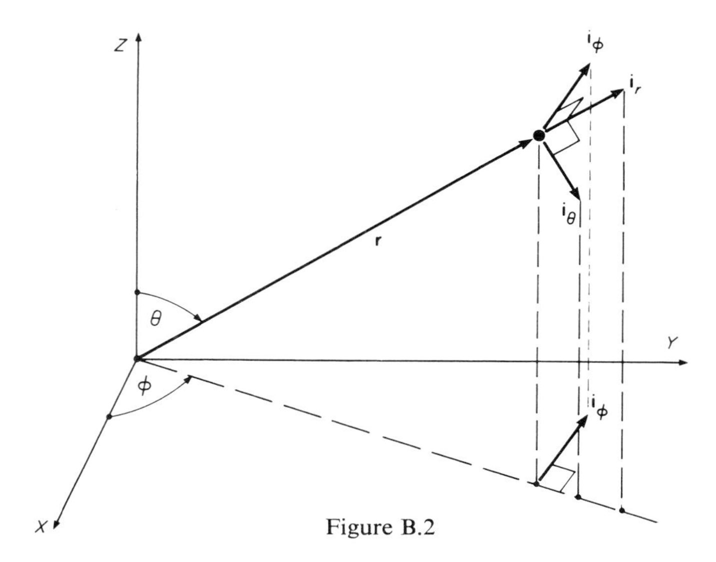


Figure B.2

In spherical polar coordinates, for a scalar quantity $u$, the gradient of $u$ (grad $u$) is

$$\nabla u = \mathbf{i}_r\frac{\partial u}{\partial r}+\mathbf{i}_\theta\frac{1}{r}\frac{\partial u}{\partial \theta}+\mathbf{i}_\phi\frac{1}{r\sin\theta}\frac{\partial u}{\partial \phi}$$

The divergence and curl of a vector **a** are given by

$$\nabla\cdot\mathbf{a} = \frac{1}{r^2}\frac{\partial}{\partial r}(r^2 a_r)+\frac{1}{r\sin\theta}\frac{\partial}{\partial\theta}(a_\theta\sin\theta)+\frac{1}{r\sin\theta}\frac{\partial a_\phi}{\partial\phi}$$

and

$$\begin{aligned}\nabla\times\mathbf{a} &= \mathbf{i}_r\frac{1}{r\sin\theta}\left\{\frac{\partial}{\partial\theta}(a_\phi\sin\theta)-\frac{\partial a_\theta}{\partial\phi}\right\}\\ &\quad+\mathbf{i}_\theta\frac{1}{r\sin\theta}\left\{\frac{\partial a_r}{\partial\phi}-\sin\theta\frac{\partial}{\partial r}(ra_\phi)\right\}\\ &\quad+\mathbf{i}_\phi\left(\frac{1}{r}\right)\left\{\frac{\partial}{\partial r}(ra_\theta)-\frac{\partial a_r}{\partial\theta}\right\}\end{aligned}$$

### The Scalar Triple Product $\mathbf{a}\cdot(\mathbf{b}\times\mathbf{C})$

The parentheses are used to show that the vector product is completed first and then the scalar product of **a** and the result of $(\mathbf{b}\times\mathbf{C})$ is taken. If **a** is equal to **b** or **C**, then the scalar triple product is zero. This must be so because $(\mathbf{b}\times\mathbf{C})$ gives a vector perpendicular to *both* **b** and **C**. If the scalar product is taken with **b** or **C**, the result has zero magnitude because both vectors in the scalar product are at right angles. Consequently, for *any* vector **C**

$$\nabla\cdot(\nabla\times\mathbf{C}) = 0$$

It can be shown that the vectors in a scalar triple product can be given a cyclic rotation without changing the scalar triple product.

$$\begin{aligned}\mathbf{a}\cdot(\mathbf{b}\times\mathbf{C}) &= \mathbf{b}\cdot(\mathbf{C}\times\mathbf{a})\\ &= \mathbf{C}\cdot(\mathbf{a}\times\mathbf{b})\\ &= \mathbf{a}\cdot(\mathbf{b}\times\mathbf{C})\end{aligned}$$

### The Vector Triple Product $\mathbf{a}\times(\mathbf{b}\times\mathbf{C})$

Again the parentheses show that the vector product $\mathbf{b}\times\mathbf{C}$ is completed first and then the result is 'cross' multiplied by **a**. The identity

$$\nabla\times(\nabla\times\mathbf{C}) = \nabla(\nabla\cdot\mathbf{C})-\nabla^2\mathbf{C}$$

can be shown to hold for *any* vector **C**.

### Vector Integrals

The proofs of the integral theorems of Stokes and Gauss are to be found in any mathematical text dealing with vector calculus and only the results are quoted here.

(i) *Gauss' Theorem*

For any vector $\mathbf{C}$

$$\int_s \mathbf{C}\cdot\mathbf{n}\,\mathrm{d}s = \int_V \nabla\cdot\mathbf{C}\,\mathrm{d}V$$

where $\mathrm{d}V$ is a volume element in the volume $V$ enclosed by the surface $s$; $\mathbf{n}$ is the outward unit normal to $s$ at an element of surface area $\mathrm{d}s$.

(ii) *Stokes' Theorem*

For any vector $\mathbf{C}$

$$\oint_\Gamma \mathbf{C}\cdot\mathrm{d}\mathbf{r} = \int_s (\nabla\times\mathbf{C})\cdot\mathbf{n}\,\mathrm{d}s$$

where $\mathrm{d}\mathbf{r}$ is an element of the closed curve $\Gamma$ which forms the boundary of the surface $s$; $\mathbf{n}$ is the unit normal to $s$ at an element of surface area $\mathrm{d}s$.

## 1(b) THE USE OF THE JACOBIAN DETERMINANT IN COORDINATE TRANSFORMATIONS

In Chapter 7 we use a Jacobian determinant to relate a volume element $\mathrm{d}V$ with coordinates $(x, y, z)$ to the corresponding volume element $\mathrm{d}V'$ in the retarded volume with coordinates $(x', y', z')$. The use of such determinants is illustrated in the following two dimensional case. The primed coordinate system $(x', y')$ is related to the $(x, y)$ system.

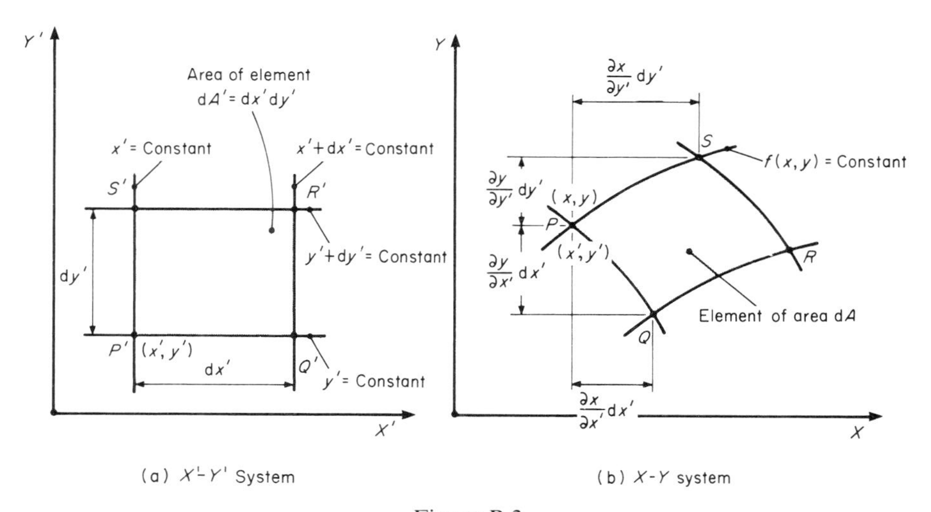


Figure B.3

We ask the question—how is the infinitesimal element of area $dx'\,dy'$ related to the corresponding element in the $(X, Y)$ system? The straight lines defining $dx'\,dy'$ have corresponding curves in the $X$-$Y$ plane, so that, for example,

$$x' = \text{constant (in the } X'\text{-}Y' \text{ plane)}$$

corresponds to

$$f(x, y) = \text{constant (in the } X\text{-}Y \text{ plane)}$$

The coordinates of $P'$ are $(x', y')$ and this point corresponds to $P$ with coordinates $(x, y)$. To move from $P'$ to $Q'$ requires a change $dx'$ in $x'$ only. Therefore we can write

$$\begin{aligned} X \text{ coordinate of } Q &= x + \text{change in } X \text{ as a result of a change } dx' \text{ in } x' \text{ only} \\ &= x + \frac{\partial x}{\partial x'}\,dx' \end{aligned}$$

$$\begin{aligned} Y \text{ coordinate of } Q &= y + \text{change in } Y \text{ as a result of a change } dx' \text{ in } x' \text{ only} \\ &= y + \frac{\partial y}{\partial x'}\,dx' \end{aligned}$$

Similarly we can write the $(X, Y)$ coordinates of the point $S$ as

$$\left(x + \frac{\partial x}{\partial y'}\,dy',\ y + \frac{\partial y}{\partial y'}\,dy'\right)$$

The area of a triangle can be expressed in terms of its coordinates and this expression can be conveniently written as a determinant. The area of the infinitesimal triangle PSQ is given by

$$\frac{1}{2}\begin{vmatrix} x & y & 1 \\ \left(x + \frac{\partial x}{\partial x'}\,dx'\right) & \left(y + \frac{\partial y}{\partial x'}\,dx'\right) & 1 \\ \left(x + \frac{\partial x}{\partial y'}\,dy'\right) & \left(y + \frac{\partial y}{\partial y'}\,dy'\right) & 1 \end{vmatrix}$$

Subtracting the first from the second and third rows leaves the determinant unchanged, and the area of the triangle is given by

$$\frac{1}{2}\begin{vmatrix} x & y & 1 \\ \frac{\partial x}{\partial x'}\,dx' & \frac{\partial y}{\partial x'}\,dx' & 0 \\ \frac{\partial x}{\partial y'}\,dy' & \frac{\partial y}{\partial y'}\,dy' & 0 \end{vmatrix} = \left(\frac{1}{2}\right) dx'\,dy' \begin{vmatrix} \frac{\partial x}{\partial x'} & \frac{\partial y}{\partial y'} \\ \frac{\partial x}{\partial y'} & \frac{\partial y}{\partial y'} \end{vmatrix}$$

The area of the rectangular element $\mathrm{d}A$ is then twice the area of this triangle

$$\mathrm{d}A = \mathrm{d}x'\,\mathrm{d}y' \begin{vmatrix} \dfrac{\partial x}{\partial x'} & \dfrac{\partial y}{\partial x'} \\ \dfrac{\partial x}{\partial y'} & \dfrac{\partial y}{\partial y'} \end{vmatrix}$$

or

$$\mathrm{d}A = \mathscr{J}\,\mathrm{d}A'$$

The three dimensional relationship

$$\mathrm{d}V = \mathscr{J}\,\mathrm{d}V'$$

is a more general result than that obtained above for two dimensions and then the Jacobian determinant is

$$\mathscr{J} = \begin{vmatrix} \dfrac{\partial x}{\partial x'} & \dfrac{\partial y}{\partial x'} & \dfrac{\partial z}{\partial x'} \\ \dfrac{\partial x}{\partial y'} & \dfrac{\partial y}{\partial y'} & \dfrac{\partial z}{\partial y'} \\ \dfrac{\partial x}{\partial z'} & \dfrac{\partial y}{\partial z'} & \dfrac{\partial z}{\partial z'} \end{vmatrix}$$

## 1(c) STANDARD INTEGRALS

$$\int_0^\pi \sin^2\theta\,\mathrm{d}\theta = \pi/2$$

$$\int_0^\pi \sin^3\theta\,\mathrm{d}\theta = 4/3$$

The time average of the square of a sinusoidal function (for example, $\sin^2 \omega t$, $\sin^2(\omega t \pm \phi)$ or $\cos^2(\omega t \pm \phi)$) is given by

$$\begin{aligned} \overline{\sin^2 \omega t} &= \frac{1}{T}\int_0^T \sin^2 \omega t\,\mathrm{d}t \\ &= \frac{\omega}{2\pi}\int_0^{2\pi/\omega} \sin^2 \omega t\,\mathrm{d}t \\ &= \tfrac{1}{2} \end{aligned}$$

where $T$ is the periodic time.

## Appendix C

# THE RELATIONSHIPS BETWEEN D AND H AND THE VECTORS E, B, P AND M

The 'electric displacement' or 'electric flux density' vector **D** is related to **E** and **P** by

$$\mathbf{D}/\varepsilon_0 = \mathbf{E}+\mathbf{P}/\varepsilon_0$$

and consequently

$$\frac{|\mathbf{D}|}{|\mathbf{E}|} = \varepsilon_0\varepsilon$$

In a magnetic material, the total current density **J** contains contributions from electric polarization currents ($\mathbf{J}_m$), conduction currents ($\mathbf{J}_c$) and currents associated with the magnetic properties of the material, $\mathbf{j}_m$. Then the Maxwell relation

$$c^2\nabla\times\mathbf{B} = \mathbf{J}/\varepsilon_0+\partial\mathbf{E}/\partial t$$

becomes

$$c^2\nabla\times\mathbf{B} = \mathbf{J}_c/\varepsilon_0+\frac{1}{\varepsilon_0}\mathbf{J}_m+\frac{1}{\varepsilon_0}\mathbf{j}_m+\partial\mathbf{E}/\partial t$$

or

$$c^2\nabla\times\mathbf{B} = \mathbf{J}_c/\varepsilon_0+\frac{1}{\varepsilon_0}\frac{\partial\mathbf{P}}{\partial t}+\frac{1}{\varepsilon_0}\nabla\times\mathbf{M}+\frac{\partial\mathbf{E}}{\partial t}$$

where the magnetization vector **M** is defined as the magnetic moment per unit volume in an analogous way to the definition of **P**. Then the equation can be written as

$$c^2\nabla\times(\mathbf{B}-\mathbf{M}/\varepsilon_0 c^2) = \mathbf{J}_c/\varepsilon_0+\frac{\partial}{\partial t}(\mathbf{E}+\mathbf{P}/\varepsilon_0)$$

or

$$\nabla\times\mathbf{H} = \mathbf{J}_c/+\partial\mathbf{D}/\partial t$$

where **H** (the magnetic field intensity) is defined by

$$\mathbf{H} = \varepsilon_0 c^2\mathbf{B}-\mathbf{M}$$

The permeability $\mu$ is then defined from

$$\mathbf{B} = \mu \mathbf{H}$$

For ferromagnetic materials $\mu$ is no longer constant, corresponding to the non-linear relationship between **D** and **E** in ferroelectric materials.

Appendix D

# LIST OF SYMBOLS

| | |
|---|---|
| $a$ | dimension (m) |
| $A$ | area ($m^2$) |
| $\mathbf{a}$ | arbitrary vector |
| $\mathbf{A}$ | magnetic vector potential ($Wb\ m^{-1}$, weber per metre) |
| $b$ | dimension (m) |
| $\mathbf{b}$ | arbitrary vector |
| $\mathbf{B}$ | magnetic field vector (T, tesla) |
| $\mathbf{B}_0$ | constant magnetic field vector (amplitude $B_0$) |
| $\mathbf{B}_0'$ | constant magnetic field vector in reflected wave (amplitude $B_0'$) |
| $\mathbf{B}_0''$ | constant magnetic field vector in transmitted wave (amplitude $B_0''$) |
| $c$ | velocity of light in free space |
| $C$ | capacitance (F, farads) |
| $\mathbf{C}$ | arbitrary vector |
| $\mathbf{D}$ | electric displacement vector ($C\ m^{-2}$) |
| e | Napierian base ($e \simeq 2.718$) |
| $\mathbf{E}$ | electric field vector ($V\ m^{-1}$) |
| $\mathbf{E}_0$ | constant electric field vector (amplitude $E_0$) |
| $\mathbf{E}_0'$ | constant electric field vector in reflected wave (amplitude $E_0'$) |
| $\mathbf{E}_0''$ | constant electric field vector in transmitted wave (amplitude $E_0''$) |
| $\mathscr{E}$ | electromagnetic energy density ($J\ m^{-3}$) |
| $f$ | frequency (Hz); function of a variable |
| $f_c$ | cut-off frequency; critical frequency (Hz) |
| $\mathbf{F}$ | force vector (N, newton) |
| $g$ | gain of an antenna; function of variable |
| $\mathbf{H}$ | magnetic field intensity vector ($A\ m^{-1}$) |
| $i$ | $\sqrt{-1}$ |
| $\mathbf{i}$ | unit vector in the positive $x$ direction |
| $\mathbf{i}_r$ | unit vector in the direction of $r$ (only) increasing |
| $\mathbf{i}_\theta$ | unit vector in the direction of $\theta$ (only) increasing |
| $\mathbf{i}_\phi$ | unit vector in the direction of $\phi$ (only) increasing |
| $\mathbf{I}$ | current vector (A, ampere) |
| $\mathbf{j}$ | unit vector in the positive $y$ direction |
| $\mathbf{j}_m$ | current density within a material giving rise to magnetization ($A\ m^{-2}$) |

| | |
|---|---|
| $\mathbf{J}$ | current density vector (A $m^{-2}$) |
| $\mathscr{J}$ | Jacobian determinant |
| $k$ | propagation constant |
| $k_x$ | $x$-component of propagation vector |
| $k_y$ | $y$-component of propagation vector |
| $k_z$ | $z$-component of propagation vector |
| $\mathbf{k}$ | unit vector in the positive $z$ direction |
| $\mathbf{K}$ | propagation vector (amplitude $K$) |
| $\mathbf{K}'$ | propagation vector for reflected wave |
| $\mathbf{K}''$ | propagation vector for transmitted wave |
| $\mathscr{k}$ | term in the expression for the refractive index |
| $l$ | length of electric dipole (m) |
| $\ell$ | term in the expression for the refractive index |
| $L$ | free charge displacement (m) |
| $m$ | mass of electric charge $q$ (kg); integer |
| $\mathbf{m}$ | magnetic dipole moment of a current loop |
| $\mathbf{M}$ | magnetization per unit volume of material (A $m^{-1}$) |
| $n$ | refractive index; integer |
| $n_i$ | imaginary part of refractive index |
| $n_r$ | real part of refractive index |
| $N$ | number of atoms, molecules or electric charges per unit volume of material |
| $\mathbf{p}$ | electric dipole moment (C m) |
| $\mathbf{p}_m$ | molecular electric dipole moment (C m) |
| $\mathbf{P}$ | electric polarization per unit volume (C $m^{-2}$) |
| $q$ | electric charge (C, coulomb) |
| $Q$ | quantity of electric charge (C) |
| $R_r$ | radiation resistance of an antenna ($\Omega$) |
| $\mathbf{r}$ | vector from reference origin to field point (m) |
| $\mathbf{r}_p$ | vector from reference origin to charge point |
| $\mathbf{r}'_p$ | vector from reference origin to retarded charge position |
| $\mathbf{R}$ | vector from charge position to field point |
| $\mathbf{R}'$ | vector from retarded charge position to field point |
| $\mathbf{R}_p$ | $\mathbf{r}_p - \mathbf{r}'_p$, vector joining retarded and present charge or current positions |
| $s$ | surface ($m^2$) |
| $\mathscr{S}$ | entropy (J $K^{-1}$ $mol^{-1}$) |
| $\mathbf{S}$ | Poynting vector (W $m^{-2}$) |
| $t$ | time (s, second) |
| $t'$ | retarded time (s, second) |
| $T$ | temperature (K, kelvin) |
| $T_C$ | Curie temperature |
| $u$ | energy (J) |
| $U$ | energy per unit volume (J $m^{-3}$) |
| $\mathbf{v}$ | velocity; phase velocity of electromagnetic wave (m $s^{-1}$) |

| | |
|---|---|
| $\mathbf{v}_{en}$ | energy velocity in an electromagnetic wave (m $s^{-1}$) |
| $\mathbf{v}'$ | retarded velocity of charge (m $s^{-1}$) |
| $V$ | volume ($m^3$) |
| $W$ | power (W, watt) |
| $W_0$ | power radiated by an antenna or free charge |
| $W_0'$ | power radiated by a bound charge |
| $W_i$ | power per unit area in a plane electromagnetic wave |
| $x, X$; $y, Y$; $z, Z$ | coordinates in rectangular system |
| $x', y', z'$ | retarded coordinates of moving electric charge |

| | |
|---|---|
| $\alpha$ (alpha) | molecular or atomic polarizability |
| $\beta$ (beta) | arbitrary constant |
| $\gamma$ (gamma) | damping constant for the motion of a charge |
| $\Gamma$ (gamma) | closed curve |
| $\delta$ (delta) | penetration depth (m) |
| $\varepsilon$ (epsilon) | relative permittivity of a material |
| $\varepsilon_0$ | absolute permittivity of free space $\varepsilon_0 \simeq 8.85 \times 10^{-12}$ F $m^{-1}$ |
| $\theta$ (theta) | angle |
| $\lambda$ (lambda) | wavelength (m) |
| $\mu$ (mu) | permeability (H $m^{-1}$) |
| $\pi$ (pi) | $\pi \simeq 3.14$ |
| $\rho$ (rho) | charge density (C $m^{-3}$) |
| $\sigma$ (sigma) | conductivity $(\Omega\ m)^{-1}$; surface charge density (C $m^{-2}$) |
| $\sigma_s$ | scattering cross section for a free charge ($m^2$) |
| $\sigma_s'$ | scattering cross section for a bound charge ($m^2$) |
| $\tau$ (tau) | inverse of damping constant $\gamma$ |
| $\phi$ (phi) | scalar potential (V) |
| $\omega$ (omega) | angular frequency ($\omega = 2\pi f$) |
| $\Omega$ (omega) | ohm |
| $\mathbf{\Omega}$ (omega) | arbitrary vector |
| $\psi$ (psi) | wave function |

## Appendix E

# NOTES ON SI UNITS

This international system is based on seven fundamental units, the metre (m) for length, the kilogramme (kg) for mass, the second (s) for time, the ampere (A) for electric current, the kelvin (K) for thermodynamic temperature, the candela (cd) for luminous intensity, the mole (mol) for the amount of substance. The derived units for the electrical quantities are:

| | | |
|---|---|---|
| Charge | coulomb | C |
| Current | ampere | A |
| Capacitance | farad | F |
| Electric potential | volt | V |
| Electric Resistance | ohm | Ω |
| Conductance | siemens | S |
| Force | newton | N |
| Energy | joule | J |
| Power | watt | W |
| Inductance | henry | H |
| Magnetic field | tesla | T |
| Electric field | $V\ m^{-1}$ | |

Multiplying factors of ten are used to bring the basic unit into a suitable range. The more commonly used factors are

| | |
|---|---|
| $10^{-15}$ | femto (f) |
| $10^{-12}$ | pico (p) |
| $10^{-9}$ | nano (n) |
| $10^{-6}$ | micro ($\mu$) |
| $10^{-3}$ | milli (m) |
| $10^{-2}$ | centi (c) |
| $10^{3}$ | kilo (k) |
| $10^{6}$ | mega (M) |
| $10^{9}$ | giga (G) |

## Appendix F

# PHYSICAL CONSTANTS

| | |
|---|---|
| $c$ (Velocity of light in free space) | $\approx 3 \times 10^{8}$ m s$^{-1}$ |
| e (Napierian base) | $\approx 2.718$ |
| $\varepsilon_0$ (Absolute permittivity of free space) | $\approx 8.85 \times 10^{-12}$ F m$^{-1}$ |
| Electronic charge | $\approx 1.6 \times 10^{-19}$ C |
| Electronic mass | $\approx 9.11 \times 10^{-31}$ kg |
| Avogadro's number | $\approx 6.02 \times 10^{26}$ (kg atomic mass)$^{-1}$ |
| Mass of hydrogen atom | $\approx 1.67 \times 10^{-27}$ kg |
| Mass of proton | $\approx 1.67 \times 10^{-27}$ kg |
| Approximate Van der Waals radius of the H atom | $\approx$ 1 Å (angstrom) $= 10^{-10}$ m |

# ANSWERS TO PROBLEMS

**Chapter 1**

1. $2.1 \times 10^{-29}$ C m.
2. $3.7 \times 10^{5}$ V m$^{-1}$; $1.4 \times 10^{7}$ V m$^{-1}$.
4. $7.8 \times 10^{-24}$ J.

**Chapter 2**

1. $1.58 \times 10^{6}$ V m$^{-1}$
2. 950 V m$^{-1}$.

**Chapter 4**

1. $\sim 10^{15}$ Hz.

**Chapter 5**

1. $6.4 \times 10^{-7}$ m.
2. 13.8 kHz; 13.8 m.
3. (a) $\sim 9.0 \times 10^{6}$ Hz; (b) $2.6 \times 10^{15}$ Hz.
4. 3 GHz.
5. $4 \times 10^{2}$ m. Energy velocity $2.598 \times 10^{8}$ m s$^{-1}$, phase velocity $3.46 \times 10^{8}$ m s$^{-1}$ at 15 GHz. These results are physically acceptable and there are no energy losses in this ideal waveguide.

**Chapter 6**

1. (a) 56.°6; (b) 53.1°.
3. Reflection energy loss in potassium bromide is $\sim 9\%$; in silver chloride the reflected loss is $\sim 21\%$.
4. $\theta_c = 41.1°$.
5. $\sim 8\%$.

## Chapter 8

1. (a) $6.6 \times 10^{-29}$ m$^2$; (b) $1.96 \times 10^{-35}$ m$^2$.
2. $8.86 \times 10^{-31}$ m$^2$.
3. 10.5.
4. 11.2 A (r.m.s. value).
5. $3 \times 10^{-2}$ V m$^{-1}$(amplitude).

# SUBJECT INDEX

Page numbers in italics indicate the start of major sections on that topic.


A

Absolute permittivity of free space, 2
Accelerated point charge, 116
Air, 101, 102
Ampère's law, 4
Angle
  of incidence, *84*
  of reflection, *84*
  of transmission, *84*
Anomalous dispersion, 52
Antenna
  electric dipole, *126*
  magnetic dipole, *129*
Atmospheric scattering of radiation, 125
Attenuation of waves, 48, 58, 68
Avogadro's number, 60

B

Boundary conditions
  dielectric, *76*
  waveguide, 66, 73
Brewster angle, 93

C

Camphor, 144
Capacitor, parallel plate, 9, 36, 38, 39
Carbon dioxide, 138
Charge
  density, 3, 31, 105
  electric, 1
Clausius–Mosotti equation, 47
Complex refractive index, 48, 58
Conducting materials, 56
Conductivity
  of metal, 57
  of seawater, 75
Cooperative effects, 144
Copper, 60
Coulomb's law, 1, 4
Critical frequency, plasma, 65, 75
Curie temperature, 146
Curie–Weiss law, 146
Curl, 151, 152
Current density, 3, 31, 105
Cut-off frequency, waveguide, 69, 73, 75

D

Damping constant, 29, 48
Damping, wave, 48
Del, 150
Dipole
  electric, 7, 30
  magnetic, 139
Dispersion
  anomalous, 52
  waveguide, 72
Displacement, electric, 156
Divergence, 150
Domain structure, 145
Drude, P. K. I., 56

E

Einstein, A., 49
Electric charge, 1
Electric dipole, 7, 30
Electric displacement, 156
Electric field, 2
Electric flux density, 156
Electric force, 1, 3
Electric susceptibility, 146
Electron, 17
  scattering cross-section, 138
Energy
  density, 17, 26, 50, 71
  in an electromagnetic field, 14, 140
  velocity, 26, 50, 72
Entropy, 141
Ethyl stearate, 144

F

Faraday's law, 4
Ferroelectric, 144

Ferromagnetic, 144
Field
  electric charge, 1
  electric dipole, 7
  local, 35, 145
  magnetic, 2
Force
  electric, 1
  Lorentz, 3
  magnetic, 2
Free-charge, energy scattered by, 121
Free space, 18

**G**

Gain, antenna, 128
Gases, 35, *43*, 143
Gauss' theorem, 5, 153
Glass, 101, 102
Gradient, 150
Grazing incidence, 97
Group velocity, 52
Guided waves, 65

**H**

Helium, 29
Hertzian dipole antenna, *126*
Hydrogen chloride, 17, 29

**I**

Incidence
  angle of, *84*
  normal, 65, 75, 93, 101
Index of refraction, 38
  complex, 48, 58
Infrared radiation, 61, 102
Inhomogeneous wave equation, 104
Integrals, standard, 155
Intensity, 94
Internal field, 35, 145
Internal reflection, total, 102
Ion number density, 65
Ionosphere, 62, 75

**J**

Jacobian transformation, 109, 153

**L**

Laser, 28
Lienard–Wiechert potentials, 108, 117, 126, 130
Liquids, 143
Local field, 35, 145
Lorentz
  force, 3, 25
  gauge, 105
  local field, 38
Lorentz, H. A., 56

**M**

Magnetic dipole, 139
Magnetic field, 2
Magnetic field intensity, 156
Magnetic force, 2
Magnetic loop antenna, 129
Magnetic permeability, 157
Magnetic pole, 4
Magnetization vector, 156
Maxwell's equations, *3*
Metals
  penetration depth, 58, 74
  refractive index, *56*
  simple model for, *56*
Methane, 29
Microwaves, 61, 74
Molecular charge, 29
Molecular dipole, 30
Molecular polarizability, 30
Molecular rotation, *143*
Moment
  electric dipole, 30, 35, 145
  magnetic dipole, 139
Monochromatic wave, 20, 44, 80

**N**

Nabla (del), 150
Non-conducting materials, 43
Non-polar gases, 47
Normal incidence, 65, 75, 93, 101

**O**

Ohm's law, 57
One dimensional wave equations, 12
Order–disorder transition, 143

**P**

Parallel plate capacitor, 9, 36, 38, 39
Penetration depth, 58, 74
Permeability, 157
Permittivity
  absolute, 2
  relative, 38

Phase
  change on reflection, *94*
  velocity, 49, 64, 68, 70, 72
Plane wave, 13, 20
Plasma, reflection from, 62, 75
Point charge, 1
Polarization
  by reflection, 93
  molecular, 30
  surface, 36
  vector, 31
Polar materials, *140*
Potassium bromide, 102
Potassium niobate, 145
Potential
  scalar, *103*
  vector, *103*
Poynting vector, 16, 26, 28, 50, 70, 121
Principle of superposition, 9, 106
Propagation constant, 44, 70, 80
  vector, 80
Proton, scattering cross-section, 138

R

Radiation
  from an antenna, *126*
  from bound charge, 124
  from free charge, *116*
  resistance, 128
Radio waves, 62, 65
Rayleigh scattering, 125
Rectangular waveguide, *65*
Reflection, 76
  from a plasma, 62, 75
Refraction, 76
Refractive complex, 48, 58
Refractive index, 38
Relative permittivity, 38
Relativity, special theory of, 49
Resistance, radiation, 128
Retarded time, 108
Rotation, molecular, *143*

S

Scalar potential, *103*
Scalar triple product, 152
Scattering of radiation
  by atmosphere, 125
  by bound charge, 124
  by free charge, 121
Seawater, 75
Silver, 74
Silver chloride, 102
SI units, 161
Skin depth, 58, 74
Snell's law, 84, 89
Solids, rotational, 144
Sommerfeld, A., 56
Special relativity, 49
Spherical polar coordinates, 151
Standard integrals, 155
Stokes' theorem, 153
Sun, 28, 65, 125
Superposition, principle of, 9, 106
Surface polarization, 36

T

TEM wave, 26, 121
TM wave, 73
Temperature dependence of relative permittivity, 143
TE wave, 72
Thomson scattering, 123
Three dimensional wave equation, 13, 19, 20, 39, 44
Time, retarded, 108
Total internal reflection, 102
Transformation, Jacobian, 109, 153
Transition, order–disorder, 143
Transmission, 48, 61
Transparent materials, 98
Transverse wave, 24, 25

U

Ultraviolet radiation, 61
Units, SI, 161

V

Vector
  potential, *103*
  Poynting, 16, 26, 28, 50, 70, 121
  relationships, *149*
Velocity
  energy, 26, 50, 72
  group, 52
  of electromagnetic waves in free space, 19, 20
  of light, 3, 19, 25, 38, 49, 72
  of propagation, 12

W

Water, 101
Wave equation
  inhomogeneous, 104
  one dimensional, 12
  three dimensional, 13, 19, 20, 39, 44
Waveguide, 61, 65, 75
Waves
  in conducting materials, *56*
  in free space, *18*
  in non-conducting materials, *43*
  velocity of, 12

X

X-rays, 61